Design and Implementation of Software Engineering for Modern Web Applications

Mohammed Moreb
Smart University College for Modern Education, Palestine

A volume in the Advances in Systems Analysis, Software Engineering, and High Performance Computing (ASASEHPC) Book Series

Published in the United States of America by
IGI Global
Engineering Science Reference (an imprint of IGI Global)
701 E. Chocolate Avenue
Hershey PA, USA 17033
Tel: 717-533-8845
Fax: 717-533-8661
E-mail: cust@igi-global.com
Web site: http://www.igi-global.com

Library of Congress Cataloging-in-Publication Data

CIP DATA PROCESSING

2024 Engineering Science Reference
ISBN(hc): 9781668448441
ISBN(sc): 9781668448458
eISBN: 9781668448465

British Cataloguing in Publication Data
A Cataloguing in Publication record for this book is available from the British Library.

Advances in Systems Analysis, Software Engineering, and High Performance Computing (ASASEHPC) Book Series

Vijayan Sugumaran
Oakland University, Rochester, USA

ISSN:2327-3453
EISSN:2327-3461

MISSION

The theory and practice of computing applications and distributed systems has emerged as one of the key areas of research driving innovations in business, engineering, and science. The fields of software engineering, systems analysis, and high performance computing offer a wide range of applications and solutions in solving computational problems for any modern organization.

The **Advances in Systems Analysis, Software Engineering, and High Performance Computing (ASASEHPC) Book Series** brings together research in the areas of distributed computing, systems and software engineering, high performance computing, and service science. This collection of publications is useful for academics, researchers, and practitioners seeking the latest practices and knowledge in this field.

Coverage

- Metadata and Semantic Web
- Network Management
- Software Engineering
- Storage Systems

IGI Global is currently accepting manuscripts for publication within this series. To submit a proposal for a volume in this series, please contact our Acquisition Editors at Acquisitions@igi-global.com or visit: http://www.igi-global.com/publish/.

The (ISSN) is published by IGI Global, 701 E. Chocolate Avenue, Hershey, PA 17033-1240, USA, www.igi-global.com. This series is composed of titles available for purchase individually; each title is edited to be contextually exclusive from any other title within the series. For pricing and ordering information please visit http://www.igi-global.com/book-series/advances-systems-analysis-software-engineering/73689. Postmaster: Send all address changes to above address.

Titles in this Series

For a list of additional titles in this series, please visit: www.igi-global.com/book-series

Revolutionizing Curricula Through Computational Thinking, Logic, and Problem Solving
Mathias Mbu Fonkam (Penn State University, USA) and Narasimha Rao Vajjhala (University of New York, Tirana, Albania)
Engineering Science Reference • copyright 2024 • 245pp • H/C (ISBN: 9798369319741) • US $245.00 (our price)

Harnessing High-Performance Computing and AI for Environmental Sustainability
Arshi Naim (King Khalid University, Saudi Arabia)
Engineering Science Reference • copyright 2024 • 401pp • H/C (ISBN: 9798369317945) • US $315.00 (our price)

Recent Trends and Future Direction for Data Analytics
Aparna Kumari (Nirma University, Ahmedabad, India)
Engineering Science Reference • copyright 2024 • 350pp • H/C (ISBN: 9798369336090) • US $345.00 (our price)

Advancing Software Engineering Through AI, Federated Learning, and Large Language Models
Avinash Kumar Sharma (Sharda University, India) Nitin Chanderwal (University of Cincinnati, USA) Amarjeet Prajapati (Jaypee Institute of Information Technology, India) Pancham Singh (Ajay Kumar Garg Engineering College, Ghaziabad, India) and Mrignainy Kansal (Ajay Kumar Garg Engineering College, Ghaziabad, India)
Engineering Science Reference • copyright 2024 • 354pp • H/C (ISBN: 9798369335024) • US $355.00 (our price)

Advancements, Applications, and Foundations of C++
Shams Al Ajrawi (Wiley Edge, USA & Alliant International University, USA) Charity Jennings (Wiley Edge, USA & University of Phoenix, USA) Paul Menefee (Wiley Edge, USA) Wathiq Mansoor (University of Dubai, UAE) and Mansoor Ahmed Alaali (Ahlia University, Bahrain)
Engineering Science Reference • copyright 2024 • 564pp • H/C (ISBN: 9798369320075) • US $295.00 (our price)

IGI Global
PUBLISHER of TIMELY KNOWLEDGE

701 East Chocolate Avenue, Hershey, PA 17033, USA
Tel: 717-533-8845 x100 • Fax: 717-533-8661
E-Mail: cust@igi-global.com • www.igi-global.com

Table of Contents

Detailed Table of Contents

This chapter provides a foundational understanding of the essential services required for hosting, managing, and securely accessing modern websites. It delves into both server-side and client-side perspectives, equipping aspiring software engineers with the knowledge and tools to design and deploy user-centric websites. Readers will explore key network concepts related to server technologies, including domain structure, domain name systems, web protocols, database servers, domain names, uniform resource locators (URLs), and web browsers. Through practical examples and explanations, the chapter emphasizes the creation of websites that deliver exceptional user experiences through innovative user interfaces.

This chapter introduces readers to the underpinning services required for software engineering for modern web applications, how to analyse, developing requirements, and modelling with classes using UML. First, the authors introduce software and software engineering to help new designers to model interactions and behaviour using UML. Then they show how to justify the difference between software and software engineering, and analysis and developing requirements for modelling using UML, which process model to use in SW process, and why.

This chapter focuses on the crucial first stage of software development: requirements specification. This stage serves as a formal agreement between the developer and the client, outlining the software system's purpose and functionality. The requirements specification provides a detailed description of the system, addressing the client's needs and expectations. It encompasses the system's capabilities (what it does and doesn't do), user interactions, quality requirements, and any design constraints. Through this process, readers will gain the knowledge and skills to select and justify appropriate tools and techniques for realizing a custom-built website during the subsequent design phase.

Chapter 4

The chapter utilizes a detailed class diagram to visually represent the system's components and their relationships, drawing upon the analysis conducted in Chapter 3. Furthermore, it explores dynamic models, including sequence diagrams, state diagrams, and activity diagrams, to capture the system's behavior and interactions over time. Finally, the chapter delves into AOES software architecture, employing UML package diagrams, component diagrams, and entity-relationship diagrams to provide a comprehensive view of the system's structure and organization.

Chapter 5

The implementation phase is the last stage of your software development, in which you will have the chance to describe any difficulties you have faced during the implementation. You must also state which parts of the requirements are met and which are not. Furthermore, the implementation part must include the user and installation guides for your system, too. Software implementation is the third stage in which you are supposed to implement the software system you have designed in the previous stage. You can use any object-oriented programming languages to implement your system, e.g., PHP and C#. In this chapter, the authors applied software engineering techniques for AOES example to implement a real case to develop a personal website with all user requirements and show in detail how they transfered all system analysis to the whole system as a real practical case.

Chapter 6

The practical example personal website development (PWS) used in this chapter can be used by any user in the world for developing their website. Users will open PWS by type the address of the site using web browser. Users will navigate to website content, and if users have credential data, they can login to admin panel. PWA will install and configure on hosting support .net technology and Microsoft platform. Admin can add any new publication or book for electronic library section by login to admin site and admin can add new users and edit user profiles. After the authors introduce PWS system, they compare it with current content management systems that the user can use it from the internet. In the next sections, they analyse the system requirements functional and nonfunctional requirements. Each of the requirements will determine how to interact with other modules using uniform modeling language (UML) to describe objectives that PWS includes.

Preface

Software Engineering is a basic concept in the digital age. The seamless operation of websites is integral to the functioning of businesses, education, government services, and personal communications. This book serves as a comprehensive guide to understanding the technologies and methodologies essential for designing, developing, and maintaining modern, secure websites. It is particularly targeted at students in senior project or capstone classes, offering a valuable companion resource to refresh their knowledge about web technologies and software development as they embark on their own projects.

As the foundation of our online interactions, a website must be meticulously crafted to provide an outstanding user experience (UX) supported by an innovative user interface (UI). The initial chapters of this book delve into the core services required to host, manage, and access a secure modern website. From domain structures and domain name systems (DNS) to web protocols, database servers, and web browsers, readers are introduced to the network concepts critical to server technologies.

The first section offers an in-depth exploration of server technologies and management services. It covers essential topics such as domain names, Uniform Resource Locators (URLs), DNS resolvers and root servers, authoritative DNS servers, domain registration, and internet protocols. Additionally, the tools for designing websites are discussed, including frontend (client-side) and backend (server-side) technologies, and the principles of UI and UX design.

The second section shifts focus to software engineering for modern web applications. This part is crucial for understanding the processes of analyzing and developing requirements, modeling with Unified Modeling Language (UML), and focusing on user tasks. Readers will gain insights into the various stages of requirements engineering, including practical project purposes, system functionalities, user descriptions, and use case specifications.

In the third section, the book addresses software design, following the requirements specification phase. It provides a clear description of the structure of a software system, detailing its components and their relationships. It covers the implementation of web applications using various programming languages. This part includes detailed guidance on developing different types of web applications, such as personal websites using C#, point of sale web applications using PHP, and online examination systems. Each chapter within this section provides step-by-step instructions for implementing specific functionalities and features.

Finally, the book concludes with a section on the deployment and testing of websites. This includes server technologies and management services for hosting secure websites, website publishing, content management, search engine optimization (SEO), and tools for ensuring good design principles. The deployment chapter guides readers through transferring website files, connecting domain names, and optimizing for search engines. The testing chapter emphasizes the importance of evaluating website performance and content to ensure it meets the desired standards.

In this era of ever-evolving digital landscapes, the ability to create, manage, and secure modern websites is an invaluable skill. This book equips readers with the knowledge and tools necessary to excel in this field, making it an essential resource for both novice and experienced web designers and developers, as well as a crucial guide for students undertaking their senior projects.

Mohammed Moreb
Smart University College for Modern Education, Palestine

Section 1
Introduction

This chapter introduces readers to the underpinning services required to host, manage and access a secure modern website before introducing and exploring the methods used by designers and developers to server-side with client side. This chapter introduce and help new designers to be able to design and deliver a site that offers an outstanding User Experience supported by an innovative User Interface, and Explain in practical the network practical concept related to server technologies associated with hosting and managing websites such as domain structure, domain name systems, web protocols, database servers, and web browsers.

Chapter 1
Basic Technology Used to Manage, Design, and Develop Modern Websites

ABSTRACT

This chapter provides a foundational understanding of the essential services required for hosting, managing, and securely accessing modern websites. It delves into both server-side and client-side perspectives, equipping aspiring software engineers with the knowledge and tools to design and deploy user-centric websites. Readers will explore key network concepts related to server technologies, including domain structure, domain name systems, web protocols, database servers, domain names, uniform resource locators (URLs), and web browsers. Through practical examples and explanations, the chapter emphasizes the creation of websites that deliver exceptional user experiences through innovative user interfaces.

BACKGROUND

This chapter introduces students to the underpinning services required to host, manage and access a secure website, before introducing and exploring the methods used by designers and developers to blend back-end technologies (server-side) with front-end technologies (client-side). This chapter helps ensure new website engineers can design and deliver a site that offers an outstanding User Experience (UX) supported by an innovative User Interface (UI), it also discusses the reasons, requirements, relationships, capabilities, and features of the systems they will be using

DOI: 10.4018/978-1-6684-4844-1.ch001

and allows them to explore various tools, techniques, and technologies with 'good design' principles to plan, design and review a multipage website (Noble, 2023).

For most users, who work in website design and development it's essential to have the practical experience to identify the purpose and types of DNS for domain names, and how it organizes and manages the protocol used, so as software engineers we need to know the basic relationship between communication protocols, server hardware, operating system, webserver software for designing, publishing and accessing the website. Evaluate the impact of common web development technologies and frameworks in terms of website design, functionality, and management.

Domain Names

Domain names are the name of the site that is built by the designer, and this name is the only one of the names of all domain names in the world that is the identity of the site. The domain name is used because the human difficult to save the huge number of Internet Protocols, for all sites of the world used by the user, so the domain name was used to facilitate the process of keeping the site in human language (Cloudflare, 2020). A DNS server is a computer server that contains a database of public IP addresses and their associated hostnames, and in most cases serves to resolve, or translate, those names to IP addresses as requested. DNS servers run special software and communicate with each other using special protocols. The domain name for the website is reserved by many sites and companies that allow us to use this service either for free or by purchasing this domain name throughout a specific time. There are many websites that offer a free domain name such as www .hyperphp.com, and www.wix.com, the registration fee of a domain name depending on the website name and the number of characters on the website, Figure 1 show hidden request that execute when register new domain in f5.com website.

Figure 1. Domain name system

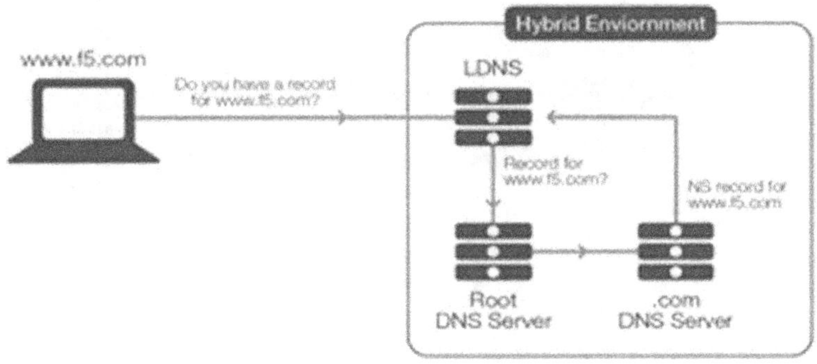

The Domain Name System, DNS for short, is likewise alluded to as the "Phone directory of the Internet". Much like searching for a name in a telephone directory to get the telephone number out, look in the DNS for a PC name to get the related IP address. The IP address is expected to associate with a server that knows just the PC name. The Domain Name System is a framework for settling PC names into IP locations and the other way around. DNS has no focal database. The data is disseminated on a huge number of name servers (DNS servers). As practice example shown in Figure 2, from step 1 when you type example.com in your internet browser until step 9 when user get the IP for the requested domain, in the event that it is your internet browser or the working framework couldn't discover the IP address in its subnet, it will send an inquiry to the following level to what is known as the resolver server for investigation. The server is fundamentally your ISP, so when the examiner Receives the question, it will inspect its impermanent memory to discover the IP for example.com, in the case he doesn't discover the domain name with IP, the inquiry will be sent to the following level is the essential server with error code.

Figure 2. Domain name operates with the protocol provider

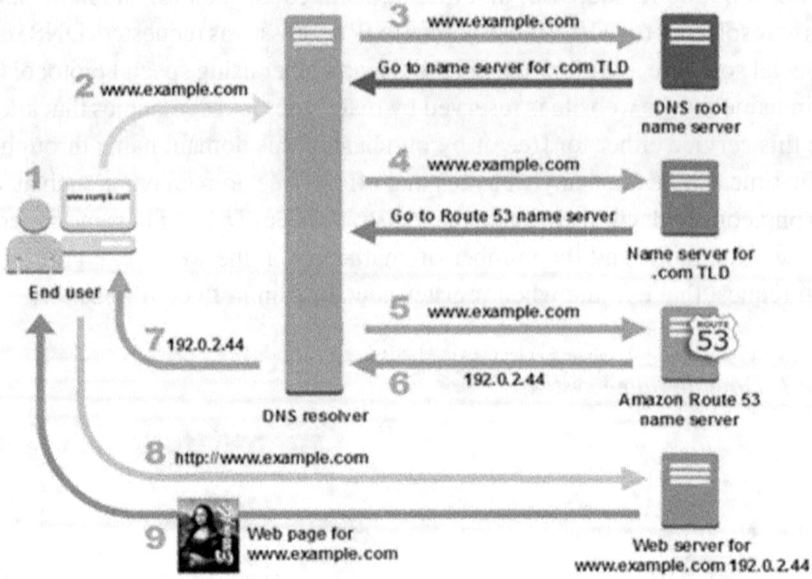

TYPE OF DNS

All DNS servers fall into one of four categories: Recursive solvers, root nameservers, TLD nameservers, and authoritative nameservers. These four DNS servers work in harmony in a standard DNS lookup (when there is no caching in play) to complete the task of transmitting the IP address for a given domain to the client (the client is typically a stub resolver-a basic resolver built into an operating system).

1. DNS Resolver

A resolver of the Domain Name System (DNS), more commonly referred to as a "DNS Lookup" tool, resolves an individual host name to an address. This form of resolution command helps to find out how to host Web services, how to support a domain name, and how to match different hardware devices with similar servers and suppliers, or with their client companies.

In some ways, by rendering the IP address of the given domain name as shown in figure 3, a DNS resolver eliminates a level of abstraction. Domain names are bits that can be accessed from anywhere on the Internet, these are identifiers for a web project that has a particular location, but that location is not defined by the domain name itself. The IP address offers much more information about where the hardware running the web project is and who owns or manages the Web space in question. For this purpose, tools such as DNS resolvers are essential for internet research and making the Web more transparent for users. For example, a user may look up or "resolve" a DNS with a specific brand name to see if the company that marked the brand owns that site or not. That form of openness can also help users to prevent other forms of Internet fraud or scams, and help to make e-commerce more available to a wider public.

Most website owners share a virtual server for each user so they can be used and handled by each webmaster owner as separate environment, as if they have complete control over the virtual server. Instead, or in addition to cloud hosting, some ISPs provide a virtual server service, the company or individual register hosted user with a website may not only have their own domain name and IP address with a virtual server, but may also manage their own file folders, attach email accounts, map addresses, and assign several domain names that resolve the primary domain name without main owner user, maintain their records, requested by another client, the resolver can circumvent the process of communicating with the nameservers, and just deliver the client the requested record from its cache.

2. DNS Root Nameserver

Every recursive resolver knows the 13 DNS root nameservers(Brownlee et al., 2001), and they are the first stop in the quest for DNS records by a recursive resolver. A root server accepts the query of a recursive resolver that contains a domain name, and the root name server responds by directing the recursive resolver to a nameserver TLD based on the domain extension (.com,.net,.org, etc.). The root nameservers are managed by a non-profit company called the Assigned Names and Numbers Network Corporation (ICANN) (McGillivray, 2014).

Notice that while there are 13 root nameservers, this does not mean that the root nameserver program includes only 13 devices. There are 13 root name server types, but there are multiple copies of each throughout the world that use Anycast routing to provide quick responses. When you added all root name server cases, you would have 632 separate servers.

DNS root servers are an integral part of the entire DNS and the Internet as shown in step 2 step 3 in figure 3, we'll introduce about what the root servers are, what they're doing, and how many are actually out there.

Root servers, or DNS root servers, are name servers that are responsible for both the DNS, and the entire internet functionality, these are the first step in any domain name resolution, which means that domain names are translated into IP addresses. Mapping names to IP addresses works with DNS zones in a progressive request. Root servers speak to the root zone, which beat the pecking order, and the root zone record is composed. The root zone file, as a result, incorporates asset records for all TLD's definitive servers. That is the reason they can work in two different ways.

- Answer legitimately to inquiries for those asset records in the root zone or
- Refer inquiries to the suitable name servers for the mentioned TLD

While in the second case they aren't legitimately engaged with name goals, root servers are in the root (allegorically and actually) of the whole Internet's framework. Without them, the vast majority of the online world we know and utilize today wouldn't be open.

Figure 3. DNS recursive resolver

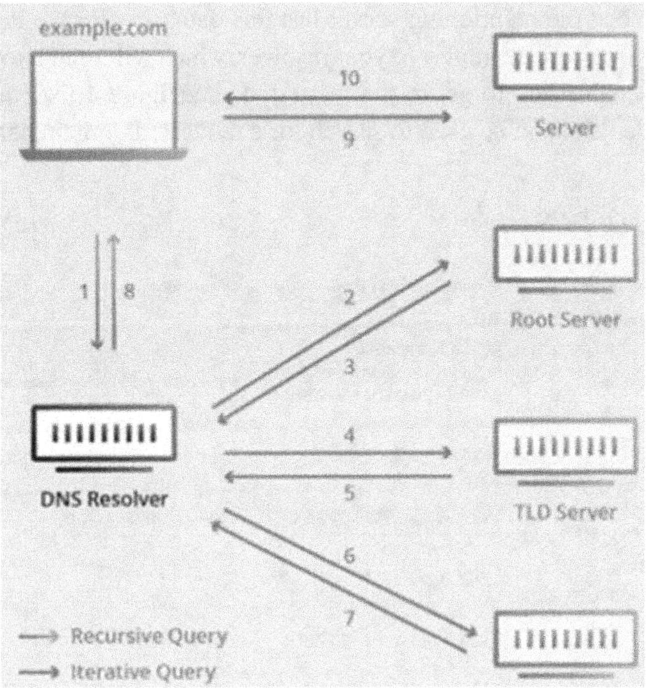

How Root Servers Work?

The way root servers work comes down to the procedure of name goals:

Step 1: When you type in www.google.com in your internet browser it will initially go to either an ISP DNS server or another DNS server you've designed, the DNS server may have the data on the space put away in reserve, and if that is the situation it will just react with the data and serve you that site.

Step 2: if DNS server can't find record for requested domain then the DNS server will send a question (request) to the root server. The root servers won't have record on a particular IP address for www.google.com, it will send request to the name servers that serve that TLD(.com).

Step 3: Root servers will restore the rundown of TLD servers so the supplier or arranged server can again send an inquiry, this opportunity to a TLD server.

Step 4: The TLD server will at that point return the definitive name server where the ideal area is put away.

Step 5: This is the point at which the server that made the solicitation sends an inquiry to the legitimate server facilitating the zone of the space being referred to.

Step 6: Once the solicitation has arrived at the definitive server, it will react to the mentioning server with the IP address for www.google.com

Step 7: When the mentioning server has this data, it will save it for future requites, and will restore the replay to your resolver, which will send it to your internet browser and enable you to get to the requested site, figure 4 lists the main DNS servers founded and can be used to search for a requested domain name.

Figure 4. List Of DNS servers

Hostname	IP address IPv4 / IPv6	Organization
a.root-servers.net	198.41.0.4, 2001:503:ba3e::2:30	VeriSign, Inc.
b.root-servers.net	199.9.14.201, 2001:500:200::b	University of Southern California (ISI)
c.root-servers.net	192.33.4.12, 2001:500:2::c	Cogent Communications
d.root-servers.net	199.7.91.13, 2001:500:2d::d	University of Maryland
e.root-servers.net	192.203.230.10, 2001:500:a8::e	NASA
f.root-servers.net	192.5.5.241, 2001:500:2f::f	Internet Systems Consortium, Inc.
g.root-servers.net	192.112.36.4, 2001:500:12::d0d	US Department of Defense (NIC)
h.root-servers.net	198.97.190.53, 2001:500:1::53	US Army (Research Lab)
i.root-servers.net	192.36.148.17, 2001:7fe::53	Netnod
j.root-servers.net	192.58.128.30, 2001:503:c27::2:30	VeriSign, Inc.
k.root-servers.net	193.0.14.129, 2001:7fd::1	RIPE NCC
l.root-servers.net	199.7.83.42, 2001:500:9f::42	ICANN
m.root-servers.net	202.12.27.33, 2001:dc3::35	WIDE Project

3. TLD Name Server?

A TLD (top-level space) is the most elevated level of area names in the root zone of the DNS of the Internet. At the end of the day the last piece of an Internet space name that pursue the last speck of a completely qualified area name. For instance, in the area name www.dnsknowledge.com, the top-level space is com.

Web Corporation for Assigned Names and Numbers (ICANN) (Weinberg, 2000) manage and configure of most top-level area. It works the Internet Assigned Numbers Authority (IANA) (McGillivray, 2014) and is liable for keeping up the DNS root zone. DNS server which keeps all root zone is called TLD name server.

A TLD nameserver holds details for all domain names that share a common domain suffix, such as .com, .net, or anything that comes after the last point in a url. For example, for every website ending in'.com, a.com TLD nameserver contains information. If a user was looking for google.com, the recursive resolver would then send a query to a.com TLD nameserver after receiving a response from a root nameserver, referring to the authoritative nameserver for that domain. The Internet Assigned Numbers Authority (IANA), a division of ICANN, operates TLD nameservers. TLD servers are divided into two main groups by the IANA: Standard top-level domains: non-country-specific domains, some of the best-known standard TLDs include.com,.org,.net,.edu, and.gov. State code top-level domains: these include any state or state-specific domains. There are examples of .uk, .us, .ru, and .jp. In addition, for infrastructure domains, there is a third category, but it is almost never used. For the. arpa domain, which was a transitional domain used to construct current DNS, this term was created; its significance today is mostly historical.

They are the highest level in the DNS structure and from here it has other types which are:

ccTLD Country Code Top Level Domains

this type is established for a geographical area for instance .cf is specified for central african republic. In order anyone to register this domain name he must be a resident in this country but, by time some of the countries now allows for registrations not only for its residents but also for other parties outside their shore to register their domains (Wilkowski, 2021).

gTLD Generic Top-Level Domain

these types just like its name anyone anywhere in the world can be registered without any restrictions these types are (Mahler, 2014):

.com stands for commercial **it** is used for the commercial organizations, now it has become the most popular and used among a lot of people for business websites.

.net stands for network, before it was specified only for institutes that worked in networks, but then just like .com it became the second popular domain and many restrictions were not applied.

.org stands for organization, was specified for non-profit websites, but also by time .org is being used also for-profit businesses.

.edu stands for education, it was made for educational institutes like schools and universities.

.gov stands for government, it was created for only Americans governmental agencies nowadays it is used for all American tribes.

.mil stands for military; this was created only for the American military branches and it still have its restrictions.

IDN ccTLD: internationalized country code top-level domains, this domain displays non-Latin characters set, means special characters.

Second-Level Domains

It is shown in next figure the TLD in the hierarchy for example, com.tr

Third-Level Domains

This is directly under the second-level domains for example, google.com.tr Consider to be a third-level domain. (techtarget, n.d.)

Figure 5. Domain names hierarchy

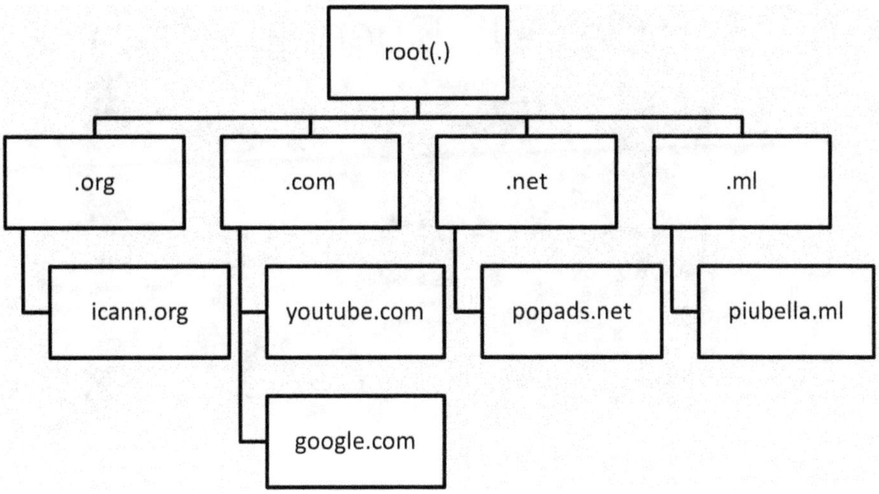

4. Authoritative Name Server?

A definitive name server gives real solution to your DNS questions, for example, mail server IP address or site IP address (An asset record). It gives unique and conclusive responses to DNS inquiries. It doesn't give just reserved answers that were gotten from another name server. Accordingly, it just returns answers to questions about domain names that are introduced previously, there are two kinds of Authoritative Name Servers (Xu et al., 2021):

a. Master Server (Essential Name Server)

A host master just makes changes to ace server zone records then each slave server gets refreshes through unique programmed refreshing instrument of the DNS convention. All slave servers keep up an indistinguishable duplicate of the ace records.

b. Slave Server (Auxiliary Name Server)

A slave server is a careful copy of ace server as show in Figure 5. auxiliary name server utilized to share DNS server load and to improve DNS zone accessibility in the event that ace server falls flat. It is prescribed that you ought to in any event have two slave servers and one ace server for every space name.

Figure 6. Slave server

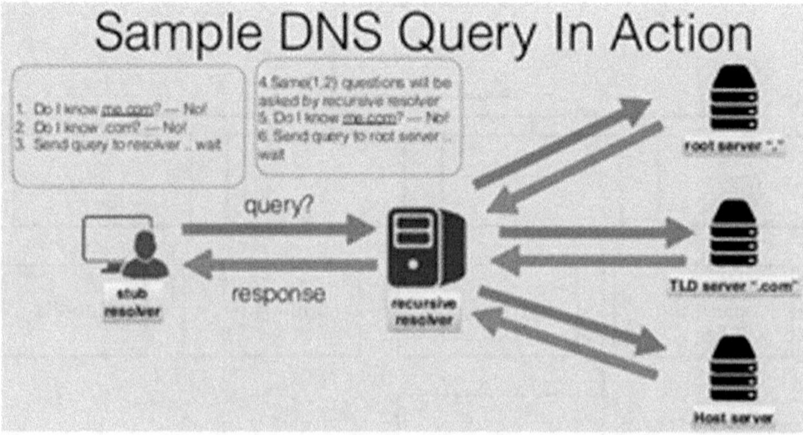

If a recursive resolver receives a reply from a TLD nameserver, that reply will direct the resolver to an authoritative nameserver. Typically, the authoritative nameserver is the last step in the search for an IP address of the resolver. The authoritative nameserver contains information unique to the domain name it represents (e.g. google.com) and can provide a recursive resolver with that server's IP address contained in the DNS A database, or if the domain has a CNAME record (prefix) it will provide an alias domain to the recursive resolver, at which point the recursive resolver will have to do a whole new DNS search to pr. Name server (often an IP-addressed A record). Cloudflare DNS distributes authoritative nameservers to make them more robust, along with anycast routing.

INTERNET PROTOCOLS

A protocol is a set of data communication rules and guidelines. Rules are defined in the communication between two or more computers for each step and process. In order to successfully transmit data, networks need to follow these rules. Compared to programming languages, protocols are structured for consistency and are based on specific rules and regulations for computationm that rule is specified in various terms and a unique name is given. Protocols define communication protocols and provide detailed information on data transmission processes. A single method can be handled concurrently by more than one protocol, such protocol alignment generates a network of protocols. Those procedures include the following (Aziz, 2020):
1: Type of task
2: Process nature

3: Data flow rate
4: Data type
5: Device management

A convention is a standard arrangement of decides that enable electronic gadgets to speak with one another. These standards incorporate what sort of information might be transmitted, what directions are utilized to send and get information, and how information moves are affirmed. You can think about a convention as a communicated in language as shown in Figure 6. Every language has its own guidelines and jargon. In the event that two individuals share a similar language, they can convey adequately. Additionally, if two equipment gadgets bolster a similar convention, they can speak with one another, paying little respect to the producer or kind of gadget. For instance, an Apple iPhone can send an email to an Android gadget utilizing a standard mail convention. A Windows PC can recive a page from a Unix-based web server utilizing a standard web convention.

Figure 7. Human protocol and computer network protocol

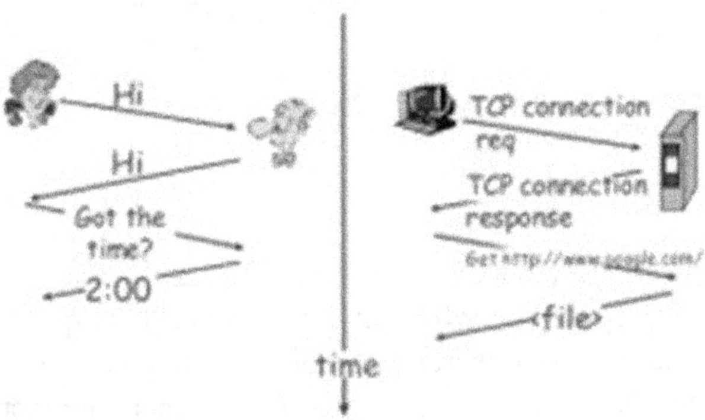

Conventions exist for a few unique applications, models incorporate wired systems administration (e.g., Ethernet), remote systems administration (e.g., 802.11ac), and Internet correspondence (e.g., IP). The Internet convention suite, which is utilized for transmitting information over the Internet, contains many conventions. These conventions might be separated into four categories (Zou et al., 2021):

Connection layer conventions build up correspondence between gadgets at an equipment level. So as to transmit information starting with one gadget then onto the next, every gadget's equipment must help a similar connection layer convention. Web layer conventions are utilized to start information moves and course them over the Internet. Transport layer conventions characterize how bundles are sent, gotten, and affirmed. Application layer conventions contain directions for explicit applications. For instance, an internet browser utilizes HTTPS to safely download the substance of a site page from a web server, an email customer utilizes SMTP to send email messages through a mail server.

Conventions are a crucial part of advanced correspondence, by and large, conventions work out of sight, so it isn't essential for run of the clients to realize how every convention function, it might be useful to acquaint yourself with some normal conventions so you can all the more likely comprehend settings in programming programs, for example, internet browsers and email customers, next we listed in figure 7 the main communications protocols, that used as internet protocols, first of all, in order to allow the website to connect with server.

Figure 8. Communication protocols

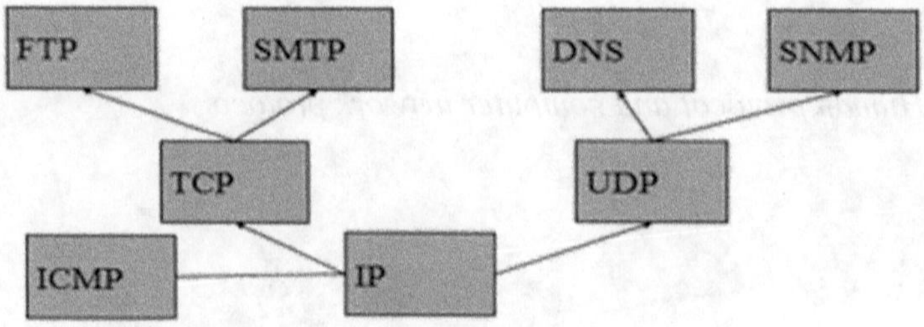

The relationship between the protocols and DNS server can be explain using practical step by step example as when you type example.com in your web browser In fact, you activate the web browser protocol called http or https, through which you can access the site, then the web browser Or the operating system try to find the IP address, if it couldn't find the IP address in its cache, it will sends a query to the next level to what is called the resolver server for analysis from your ISP server, so when the analyst receives the query, it will save in temporary memory to find the IP address for example.com, and if he doesn't find it, the query will be sent to The next level is the basic server, the primary servers are the top or root of the DNS hierarchy. There are 13 sets of these basic servers are strategically placed around the world, which operated by 12 different organizations and each group of these basic servers have their own IP address. So, when the server receives the core query the IP address of example.com, the primary server will not recognize What is the IP address, but the primary server knows where to send Resolver to find the IP address. So, the root server will direct parser to the TLD or the top-level domain server for the domain dot com. So, the analyst will now ask the TLD server for the example.com IP address. The top-level domain server stores address for top-level information domains, such as.com and.net dot org and etc. The TLD server specifically controls the dot com domain that example.com is part of. So, when the TLD server receives the query for the example.com IP address, The TLD server will not recognize the example.com IP addresses, therefore the TLD will direct the analyst to the next and last level, which is official name servers. the official server or nameservers are responsible for knowing everything about the range that includes the IP address which is the final authority, and therefore when the official name server receives the query from the analyst, the name server will respond with the example.com IP address. Finally, the analyst will tell your computer about the IP address of example.com, your computer can now browse the example web page It is important to note that once the resolver receives the IP address, you'll It stores it in its cache in case another query is received to example.com for it you don't need to go through all those steps again. Next, we listed the main protocols that used by website and as software engineer need to know where to use and why as following (Kumar & Roy, 2021; Oliveira et al., 2019):

HTTP: hypertext transfer protocol, it establishes the connection between the client and the server but this protocol is not secure.

SSL: secure sockets layer, used to encrypt the connection between the client and the web server over the internet. SSL is the protocol used for majority of secure transactions on the Internet. Encrypting/decrypting with public/private keys are used during the handshake and the session key used after that, main step for establish the secure connection using SSL as the next steps:

1. Browser connects to a web server (website) secured with SSL (https). Browser requests that the server identify itself.
2. Server sends its SSL Certificate, including the server's public key.
3. Browser checks the certificate a list of trusted CAs and sends back a symmetric session key using the server's public key.
4. Server decrypts the symmetric session key using its private key and sends back an acknowledgement encrypted with the session key to start the encrypted session.
5. Server and Browser now encrypt all transmitted data with the session key.

SSH: is a relatively simple protocol creates a "secure tunnel", Insecure command sent thru SSH tunnel are then secure, and can be used with command like *rlogin*, many tools can use to prevent the attack in SSH such as:
§ An X.509 certificate (as in Tectia SSH and SSL/TLS)
§ Some kind of proprietary certificate mechanism (e.g., OpenSSH)
§ A public key on the client and a private key on the server (e.g., SSH)
§ A shared secret value (e.g., IPSec with preshared keys).

HTTPS: hypertext transfer protocol secure, it works the same as HTTP but in this one my data will be secured actually it is a combination of HTTP and SSL protocols as shown in Figure 8.

Figure 9. http vs. https

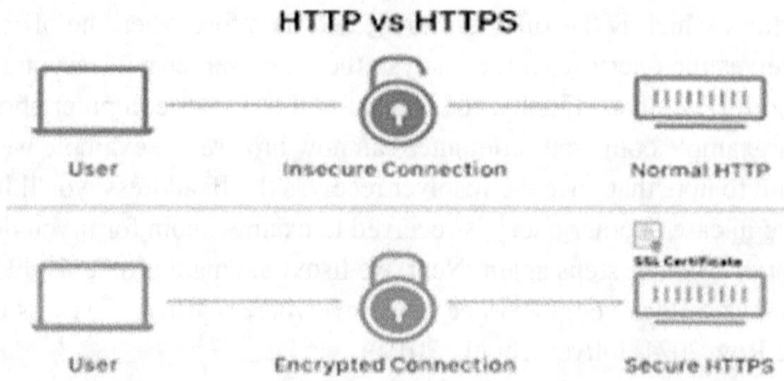

IP: internet protocol, it is a set of instructions used to route and address the packets in order to get to the destination by internet.

TCP: stands for transmission control protocol, this protocol is responsible from transmitting messages sent from the source to its destination divided in packets.

UDP: stands for user datagram protocol, this one also looks like TCP but in UDP sending of the packets is continually without establishing a communication first with the destination, it is faster compared to TCP.

Figure 10. TCP/UDP

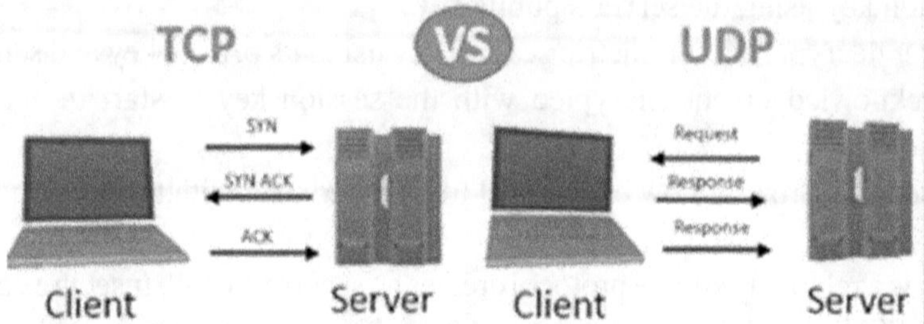

POP: Stands for post office protocol, and it is responsible for only receive the message and download it in the user's inbox, so it is the incoming server, it takes the email from remote mail servers and gives it back to the user. It belongs to the application layer in OSI model. This protocol is also called a message access agent.

SMTP: simple mail transport protocol, its main job is to send emails from the client to the receiver's mail servers, so in simple words it is a message transfer agent as shown in Figure 10. It also belongs to the application layer in OSI model.

Figure 11. POP/SMTP

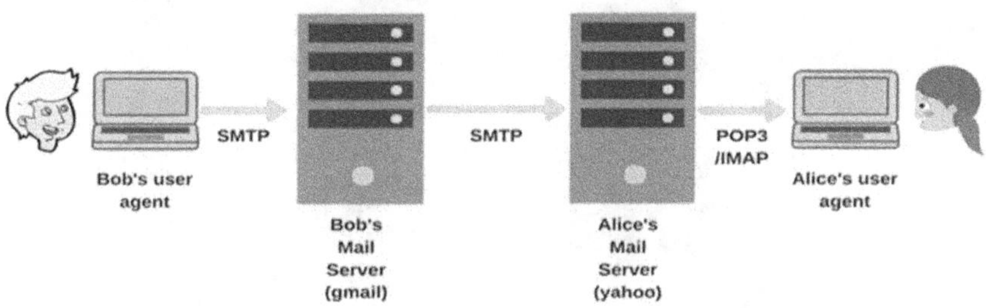

FTP: file transfer protocol uses port 21 by default to transfer files between two machines, could be media, text files, program files or documents, it transfers data from the user's computer to a web server for example.

SNMP: Stands for simple network management protocol, it belongs to the application layer in OSI model, it is used to manage and observe devices that are combined in a network and it is all functions.

ICMP: Stands for internet control message protocol, it is used also in networked devices to report any problem relates to routers for example an IP protocol cannot show messages when he cannot reach a specific website or a service instead it uses the ICMP protocol to show messages like, this service is not available now, or this router could not be reached.

Web Server (Hosting Server)

Web server or hosting server from its name we can conclude that it serves data, it uses HTTP or HTTPS to serve the requests files from browser, in responding to the client's request, means the web server stores our web files and responds to the client's entered query by the TCP/IP protocol.

A web server is a program that utilizations HTTP (Hypertext Transfer Protocol) to serve documents that requested by clients' as website pages, in light of their solicitations, and are diverted by HTTP customers to their PCs, Custom PCs and gadgets may likewise be alluded to as Web servers.

Figure 12. Web server

This procedure is a case of a customer/server model, all PCs that host Web destinations must have Web server programming. Driving web servers to incorporate Apache (the most generally introduced web server), Microsoft Internet Information Server (IIS) and nginx (Clear Engine X) from NGNIX. Other web servers incorporate Novell's NetWare, Google Web Server (GWS) and a group of IBM Domino servers.

Web servers frequently come as a major aspect of a bigger bundle of Internet and intranet programs for email administration, downloading FTP record demands, and making distributing website pages. Contemplations for picking a web server incorporate its adequacy with the working framework and different servers, its capacity to deal with server-side programming, security highlights, organization apparatuses, web crawler, and related webpage building instruments.

Figure 13. Web server-browser

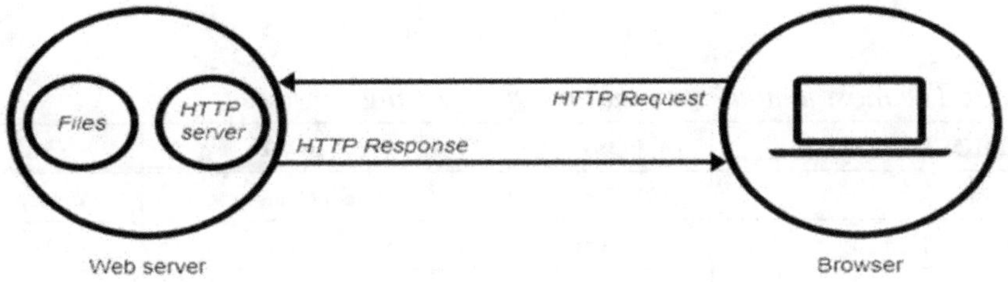

A hosting server characterized by very high specifications the processor for the speed of data transfer between the device and the user who requests the website and contains a very high random storage memory and be located within a dedicated aerial environment, with a high data transmission cable within a fire-protected center. There are many company investigate in hosting service and allow user to use their server for fee or by registration, this can be either free or paid annually depending on the company providing those services as theirs many company provide this service free for training and limited use such as dot.tk, hover, godady, domain, hostgator, and bytehose as listed in table 1. A hosting server consists of:

- **Hardware:** we need the server hardware to be connected to the internet, it also has its (hard disk drive, motherboard, CPU, RAM etc.).
- **Software:** it defines the operating system, could be (Linux, windows), and web server might be (apache, windows IIS, IBM websphere).

The most important types of hosting can be booked within the website, game or application, and below we listed the five types of hosting server as follow:

a. SHARED

It is a huge server but be shared between you and anyone else who has participated with you in that server.

b. DEDICATED

It is a server for one person only, but you can sell part of it to another person if you want.

c. VPS

It is also a special server, but it is imaginary is not real of course can be used normally.

d. CLOUD

It is a server but more dedicated to large companies because it needs very high storage space in order to meet all requests of visitors to the site.

e. MANAGED

A server is possible to be any of the servers that mentioned previously, but more expensive.

Table 1. The most famous sites that provide hosting services

WEB HOST	COST	WEB HOST	Cost
Bluehost	$2.75/mo	A Small Orange	$5.92/mo
HostGator Cloud	$2.99/mo	Site5 Hosting	$6.95/mo
Hostinger	$0.99/mo	JustHost	$3.95/mo
SiteGround	$3.95/mo	MDDHosting	$4.00/mo
GoDaddy	$2.99/mo	HostPapa	$1.00/mo
Web Hosting Buzz	$4.99/mo	HostGator (Shared)	$2.75/mo
A2 Hosting	$3.92/mo	Hosting24	$1.95/mo
iPage	$1.99/mo	FastComet	$2.95/mo
DreamHost	$2.59/mo	LunarPages	$3.95/mo
Midphase	$3.95/mo	One.com	$2.45/mo
WestHost	$1.99/mo	HostNine	$6.99/mo
Web Hosting Hub	$4.99/mo	HostRocket	$5.99/mo
NameCheap	$1.28/mo	HostMetro	$2.95/mo
HostMonster	$4.95/mo	WebHostingPad	$1.99/mo

Table 1 listed as example of famous company that provide hosting services related to the hosting varies depending on the type and according to the features provided by that company for this hosting, the most important types of hosting such as Web Host, Host application, and Game Host, next we introduce the main server that used in website development and design with purpose to use that should be families for you as software engineering, they specials in website developments, as following:

1. Proxy Server

Figure 14. Proxy server

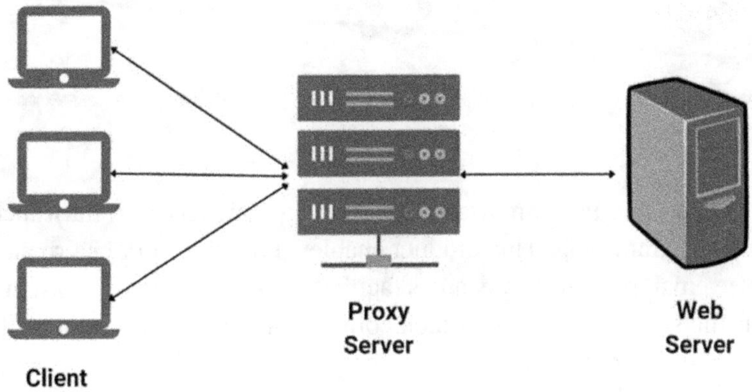

The proxy server goes about as a middle person or door between the client and the Internet (the destinations they peruse). Intermediary servers offer a few degrees of protection, usefulness and security, contingent upon client request or organization approach. When utilizing an intermediary server, Internet traffic is experiencing the intermediary server on the way to the location that the client mentioned. The solicitation at that point returns through a similar intermediary server (there are special cases to this standard), and the intermediary server advances the information from the site to you (Firmansyah & Riadi, 2014).

Current intermediary servers accomplish more than divert web demands, all for the sake of information security and system execution. Intermediary servers go about as a firewall and web channel, giving shared system associations and reserve information to accelerate basic solicitations. A decent intermediary server keeps clients and the inward system shielded from terrible contents found in the wild Internet at some level this server can give a significant level of security.

2. Mail Server

A mail server (or email server) is a PC framework that sends and gets email. By and large, web servers and mail servers are consolidated in a solitary machine. Be that as it may, enormous ISPs and open email administrations, (for example, Gmail and Hotmail) may utilize devoted equipment for sending and accepting email.

Figure 15. Mail server

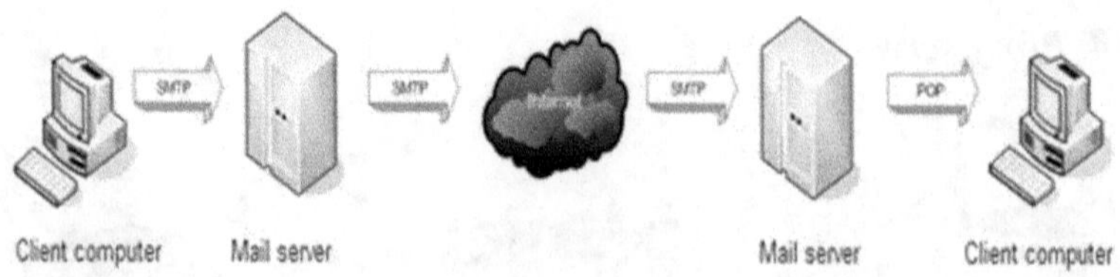

All together for a PC framework to work as a mail server, it must incorporate mail server programming. This product enables the framework chairman to make and oversee email represents any areas facilitated on the server. For instance, if the server has the space name "abchorizon.com," it can give email accounts finishing off with "@abchorizon.com."

Mail servers send and get email utilizing standard email conventions. For instance, the SMTP convention sends messages and handles active mail demands. The IMAP and POP3 conventions get messages and are utilized to process approaching mail. At the point when you sign on to a mail server utilizing a webmail interface or email customer, these conventions handle every one of the associations in the background.

Mail server programming is accessible for numerous stages. The most well-known mail server for Windows is Microsoft Exchange Server, an endeavor item utilized by huge organizations. In any case, numerous different choices exist, including Ipswitch IMail Server, IceWarp Mail Server, MailEnable, and hMailServer. Famous Linux alternatives incorporate Exim for sending letters and Dovecot and Courier for getting mail.

3. Platform Server

An assortment of servers that contain equipment and working framework. For instance, Windows Server System alludes to x86 CPUs and a few forms of Windows Server, for example, Windows Server 2022. In any case, the expression "Linux stage" is somewhat ambiguous. Albeit most Linux servers are x86-based, Linux work with practically all equipment stages. Rather than the customer stage (Barrett et al., 2020).

4. Application server

Application Server is a sort of server that is intended to introduce, run, host, and administrations related with clients, and to give numerous IT and endeavor administrations. It offers ease in facilitating and exhibiting propelled client or business applications, which are utilized by numerous clients living outside the district or living in a similar region at the same time (Xian et al., 2008).

An application server comprises of a server working framework (OS) and servers that work together to give the best figuring escalated activities and administrations for an application. The most significant highlights of an application server remember repetition for high accessibility, information, client the executives, load adjusting, information/application security, and a unified administration interface. The application server can likewise be associated by means of big business, system or intranet frameworks and can be gotten to remotely by means of the Internet.

Figure 16. Application server

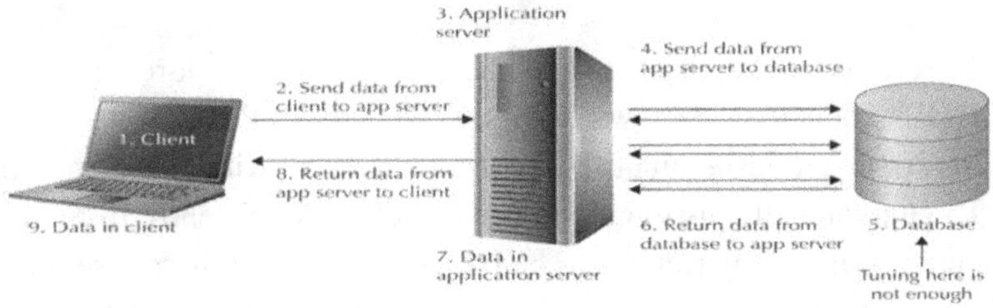

Contingent upon the introduced application, the application server might be classified in a few different ways, including a web server, a database application server, a broadly useful application server, or an undertaking application server (EA).

5. Real-Time Communication Server

Figure 17. Real-time communication server

Real-time communication server (RTC) is a word utilized when alluding to live interchanges that happen immediately in transmission time, RTC information and messages are not put away among send and get, it's commonly a distributed transmission, as opposed to communicate or multicast, and information move modes are as per the following: Unidirectional: Bidirectional, on a solitary bearer or circuit Full duplex: Bidirectional and concurrent happens on a solitary holder or circuit Instances of RTC are Internet, landlines, portable/phones, texting (IM), Internet hand-off visit, video conferencing, remotely coordinating, and mechanized telepresence. Messages, notice loads up, and writes are not RTC directs yet happen in time-evolving mode, where there is a critical deferral between information move and gathering.

RTC highlights were first presented in windows XP, and included Microsoft Office Communicator, MSN Messenger, Windows Messenger, continuous sound and video, and texting. Microsoft working frameworks and programming applications incorporate RTC stages that are made out of segment bundles that help RTC.

6. FTP Server

A FTP server is a PC with a File Transfer Protocol (FTP) address that is planned to get a FTP association as shown in figure 16. The FTP server needs a TCP/IP system to work and relies upon utilizing devoted servers with at least one FTP customers. To guarantee that associations can be set up consistently from customers, the FTP server is normally running. The FTP server is a significant segment of the FTP design and trades records over the Internet. When client request FTP ip server, the user set up an association with the FTP server, the client's name and secret key are sent utilizing user and pass value, if credentials is correct then the FTP server is acknowledged, an affirmation is sent through the customer and the session can be begun. In the event that a FTP association is accessible, it is conceivable to continue the download on the off chance that it has not been effectively finished already, the FTP server enables you to download and transfer records. There might be get to limitations as controlled by the FTP server chairman to download various documents and from various organizers on the FTP server, the FTP server can give an association with clients without requiring login accreditations; records on FTP servers can be recovered by well-known internet browsers, yet may not bolster convention augmentations, for example, FTPS. All FTP site tends to begin with ftp:/ (Wulandari et al., 2022).

Figure 18. FTP server

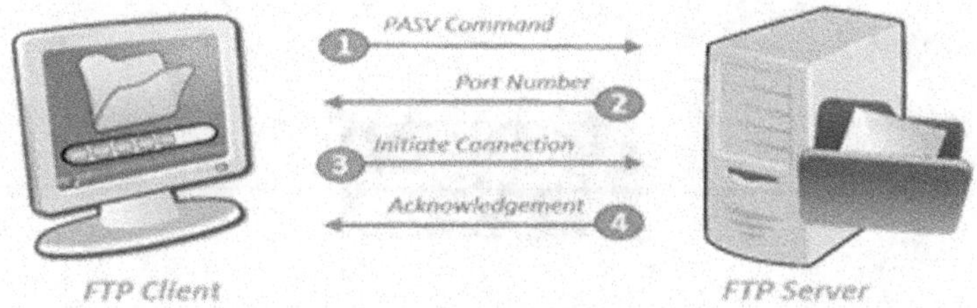

7. Collaboration Server

Figure 19. Collaboration server

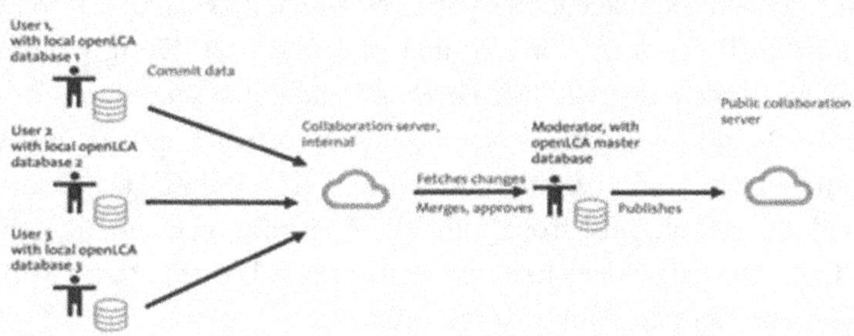

From various perspectives, joint effort programming, once called 'groupware,' shows the first intensity of the Web. Coordinated effort programming intended to empower clients to team up, paying little heed to area, by means of the Internet or a corporate intranet and to cooperate in a virtual air (Bell et al., 2014).

8. Telnet Server

Figure 20. Telnet server

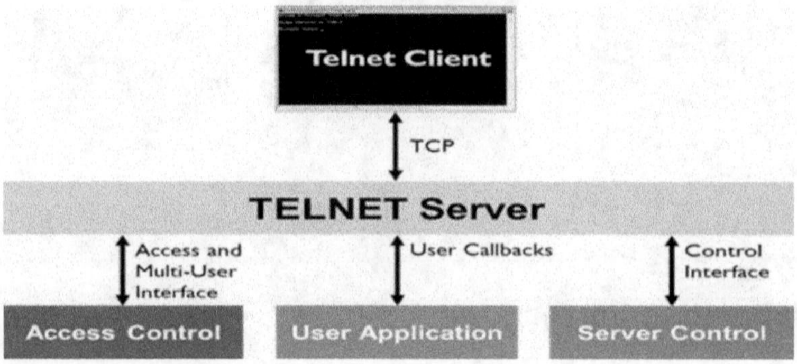

Telnet is a convention that enables you to convey long separations to PCs (called has) over a TCP/IP arrange, (for example, the Internet). By utilizing the telnet customer programming on your PC, you can make an association with the telnet server (that is, the remote host). When a telnet customer builds up an association with the remote host, your customer turns into a virtual machine, enabling you to interface with the remote host from your PC. As a rule, you should sign on to the remote host, which necessitates that you have a record on this framework, you can sign in as a visitor or open without a record.

9. Open Source Server

From your basic open-source server working framework to the server programming that encourages you complete your activity, open-source programming is a basic piece of numerous IT foundations, open-source alludes to any program whose source code is made accessible for use or alteration as clients or different engineers see fit. Open-source programming is typically created as an open coordinated effort and made uninhibitedly accessible (Garais & Enaceanu, 2016).

10. Virtual Server

Figure 21. Virtual server

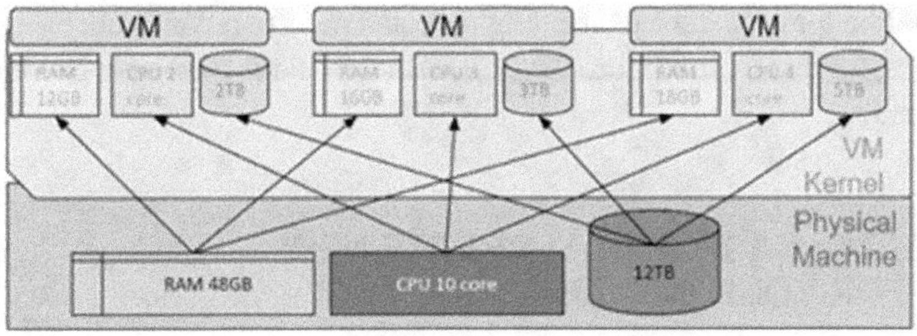

A virtual server as illustrated in Figure 19 is a server (a computer and many application programs on it) that include set of virtual server install on one physical server, but each virtual can include separate environment. Most website owners share this virtual server so that it can be used and handled by each webmaster owner as if they have complete control over the virtual server. Instead, or in addition to cloud hosting, some ISPs provide a virtual server service, the company or individual with a website may not only have their own domain name and IP address with a virtual server, but may also manage their own file folders, attach email accounts, map addresses, and assign several domain names that resolve the primary domain name without ISP intervention, maintain their records, Handle documents, evaluate data, and manage passwords. Virtual server users, however, do not need to handle server activity hardware aspects and share the cost of costly line connections to the Internet effectively (Christanto & Sani Suprayogi, 2018).

11. DHCP Server

Figure 22. DHCP server

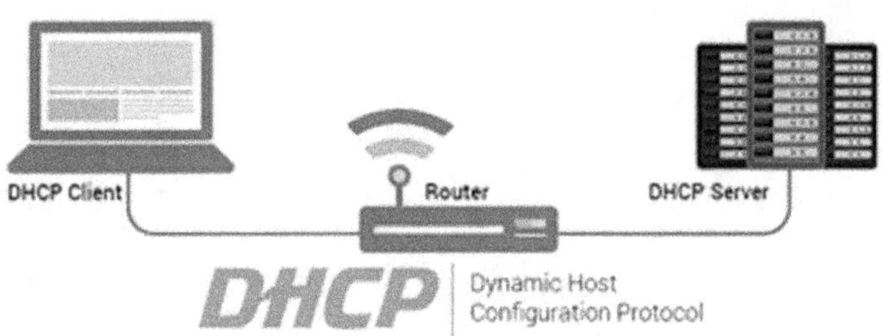

DHCP server represents dynamic host setup convention (Syarifudin, 2020), it will probably appoint interesting IP delivers to our hosts it additionally gives other system tends to, for example, in our office we have 2 pcs, a switch and a Rooter, so to connect our PCs with internet or network they need an IP address, manual configuration for each device take long time and need to reconfigure each time the ip changes, for that we configure the PC to use DHCP for automatic get the ip from DHCP server.

PRACTICAL CASE STUDY: CREATE NEW NAME FOR YOUR DOMAIN

Now there are many ways to get your domain registered like Domian.com, GoDaddy.com as example, in our practical example we selected one company that offer domain registration, and theirs's many available websites but most of them offer the registration with fees, and other site we listed next give this service as free for limited time, below we listed practical example for free registration domain for your sites, main step as follow:

Step 1: We need to go to the website www.freenom.com and add your domain in the search area, then check in available until find suitable domain name, such as soft4call or your own domain you need to register.

Step 2: Choose the type of selected domain (for training use free domain), as shown in Figure 21, then click checkout.

Figure 23. Select new domain

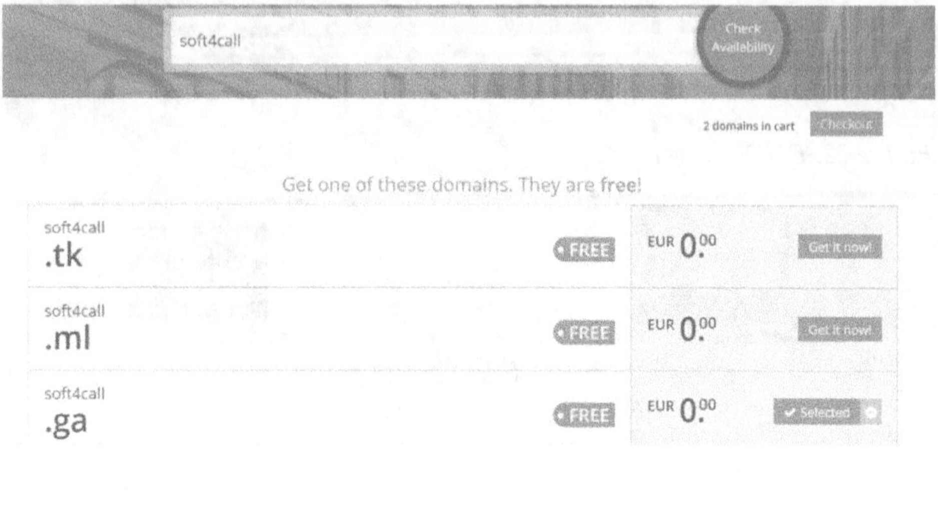

Step 3: We need to press Checkout to pick the details of the domain.

Figure 24. 3-step domain

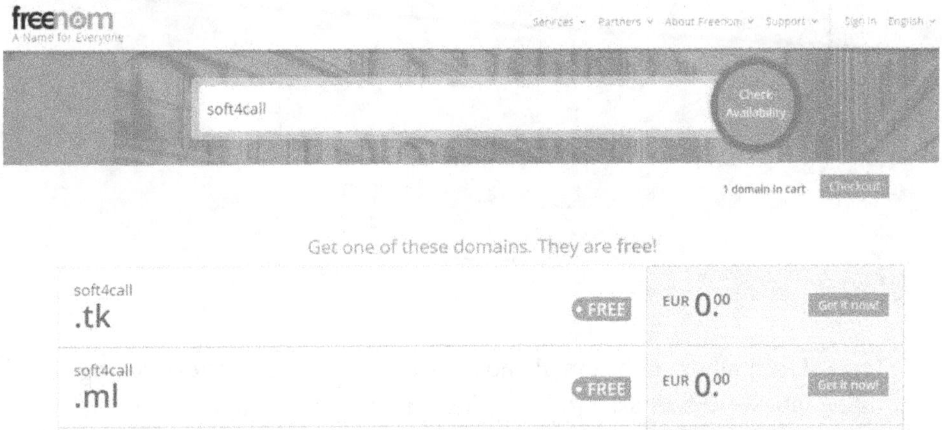

Step 4: We need to choose the domain's validity period, select the period of time that you would like for selected domain, for our case we recommend reader to chose 12 months since it is the longest free period that freenom is offering as shown in Figure 25.

Figure 25. Period of domain

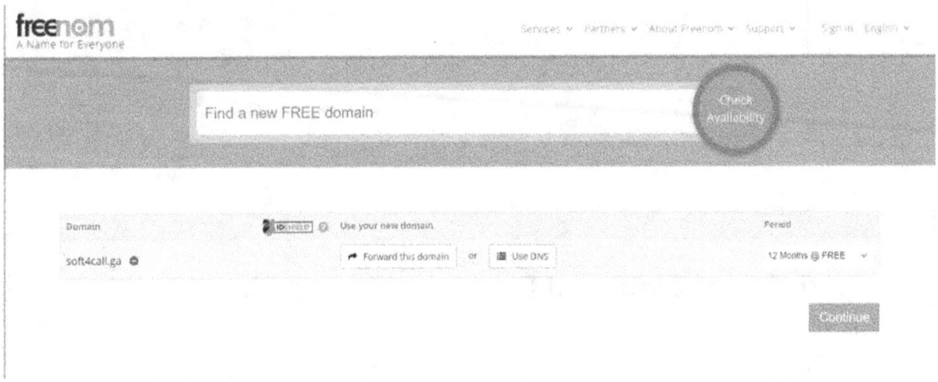

Step 5: To receive an email, we need to click on enter different email, as shown in Figure 26.

Figure 26. Review and checkout domain

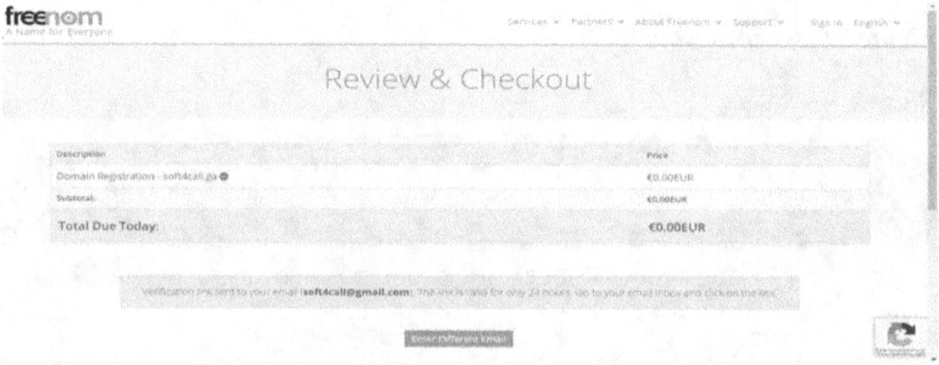

Step 6: Open inbox of the email and click on the link sent, after clicking on the link in the inbox, we will aske to relogging for activation, and as shown in Figure 27 the registered domain is ready to use.

Figure 27. 8-step domain

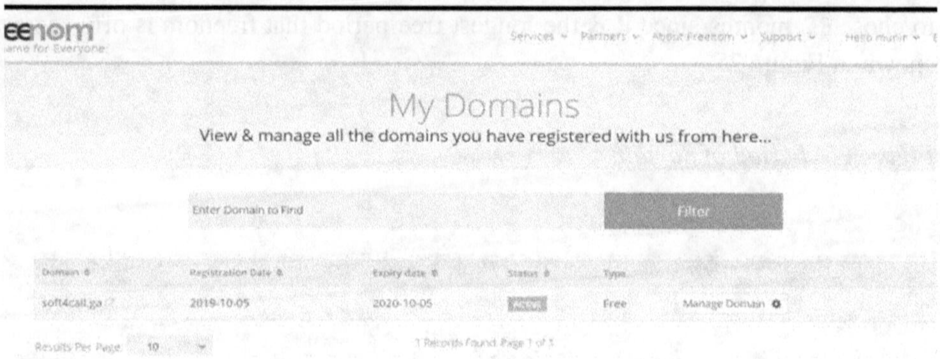

Frontend (Client-Side) and Backend (Server-Side)

Front-End

The front-end is everything engaged with what the client sees, including structure and a few dialects like HTML and CSS (Williams et al., 2019). There are a variety of employments related with the front-end. Remember that a great deal of these titles is emotional, so while front-end engineer may mean something at one organization, it can mean something totally unique at another organization.

A static site is something like a site for an eatery or hair salon. It doesn't require any data to be put away in a database. The pages will quite often remain the equivalent, except if it's the ideal opportunity for an upgrade. A front-end engineer might be required to have a grip on testing, just as be knowledgeable in HTML, CSS and JavaScript. This individual could possibly have involvement in making the plan in a structure program. An alternate variant of this title is front-end engineer. Individuals who work with explicit front-end dialects like JavaScript Developer are additionally viewed as front-end designers (Dziallas, 2020).

Figure 28. Front-end/back-end

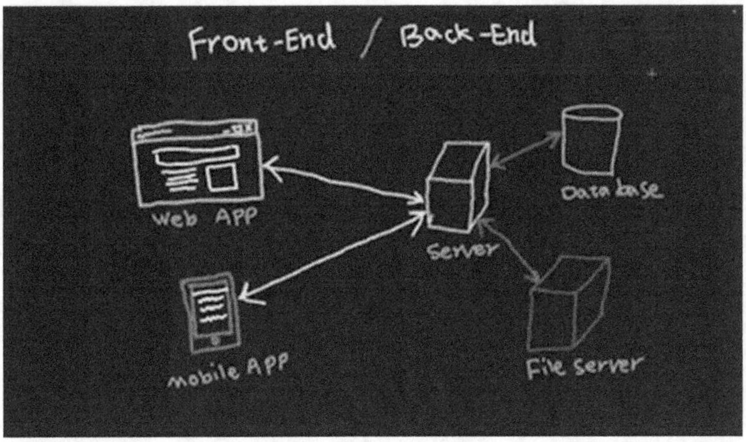

Back-End

The back-end, or the "server-side", is essentially how the site works, updates, and changes. This alludes to everything the client can't find in the program, similar to databases and servers. Normally individuals who take a shot at the back-end are gotten back to end software engineers or back-end designers. back-end Developers are for the most part stressed over things like security, structure, and substance the board. They typically know and can utilize dialects like HTML and CSS, however it's not their core interest (Raja et al., 2018).

Making a powerful site requires back-end designers, or if nothing else back-end advancement. A unique site is a site that is always showing signs of change and refreshed continuously. Most locales are dynamic destinations, instead of static destinations. Facebook, Google Maps and this blog are altogether viewed as unique destinations since their substance is continually changing and refreshing.

A unique site requires a database to work appropriately. All data is put away in the database, similar to client profiles or pictures they've transferred, or blog entries. Web engineers work with programming dialects like PHP or .Net, since they have to work with something the database gets it. The code they compose speaks with the server and afterward mentions to the program what to use from the database.

UI (User Interface)

Designer is essentially a visual planner and is commonly centered around structure. They're not generally engaged with the usage of the plan, yet they may know light HTML and CSS so they can impart their thoughts all the more adequately to the front-end designers. Just like we mentioned back in frontend and how it is everything that the user can see and interact with, it is called a user interface, it is all about selecting colors, backgrounds, and buttons, it is really important in web designing because it determine whether the user stays or leave the website, so every website must have a different and unique user interfaces according to what the website, if it is commercial it must be strongly developed by programmers in a friendly, unique attracting way, that will let clients to navigate it easily and visit it constantly, and in that way that brand will maintain its signature style.

UX (User Experience)

The user experience relates to the experiences between a user and a specific brand, application or an operating system, the better the UX is the more easy it attracts the visitors, for example even considering the first visitors to a website if they can't find a way to sign up, or get to contact, or viewing the products then that is a big problem for the UX, no matter how complicated the website is, it must easily guide the user throughout the website. If there is a website that is really looking great but it is difficult to use then here, we say that it has a perfect UI but a very poor UX. Creators work in the front-end, examining and inquiring about how individuals utilize the destinations. At that point they make changes through a ton of testing. •A Front-End Developer or Designer can make a site with no back-end advancement. The website they would make without a web engineer, or utilizing the back-end, is a static webpage.

Figure 29. UX vs. UI

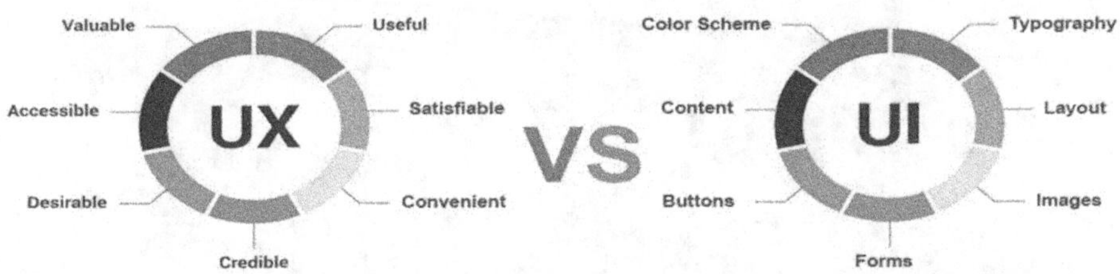

Full Stack

developers are experts in every stage of the web development process to qualify as a full stack developer you need to be proficient in both front-end and back-end languages and frameworks as well as in server network and hosting environments it also be expected to understand business logic and user experience we do distinguish between front down and back-end but it's important to have a grounding in both these days is not so much a question of should I learn front-end or back-end development most employers will expect you to be comfortable in each area if you're new to the field make sure you build up a solid overview of the entire development process and have a clear understanding on how everything works once you've got this foundation in place you can drill down to specifics and decide where to specialize further down the line no matter where you are at in your career you'll always work closely with both front down them back-end developers get to grips with both (Baiskar, 2022; Petrikoglou & Kaskalis, 2019; Zammetti, 2020).

Figure 30. OSI layers

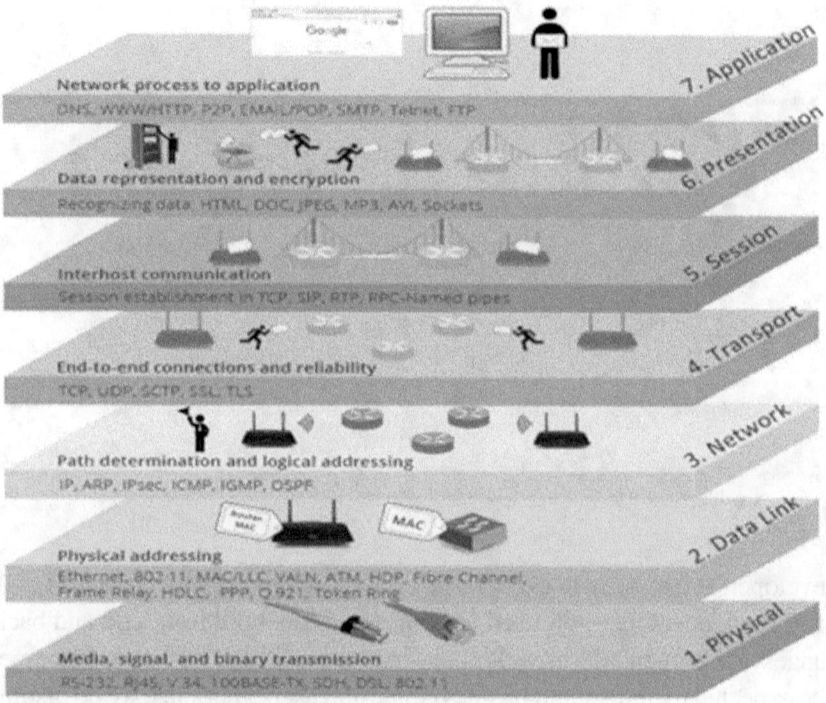

Open System Interconnection Model

OSI Model defines and is used to understand How data is transferred from one computer to another in a computer network? In the most basic form, two computers (end-points) connected to each other with LAN Cable (Network Media) and connectors (RJ45) sharing data with the help of NICs forms a network. But if one computer is based on Microsoft Windows and the other one has MAC OS installed, then how these two computers are going to communicate with each other? In order to accomplish successful communication b/w computers or networks of different architectures, 7-layered OSI Model or Open System Interconnection Model was introduced by International Organization for Standardization in 1984 containing: Application Layer, Presentation Layer, Session Layer, Transport Layer, Network Layer, Data Link Layer, Physical Layer.Note that each layer is a package of protocols (set of rules and conventions). If I say, Application Layer, then it does not mean that it includes computer applications like Chrome, Firefox etc. But it includes application layer protocols that are needed to make these applications work correctly in a network or Internet (Zimmermann, 1980).

The application layer is used by network applications. Network application means computer applications that use internet like Google Chrome, Firefox, Outlook, Skype etc. Web Browser is a network application running in your PC. It does not reside in the Application Layer. But it uses application layer protocols - HTTP or HTTPS - to do web surfing. Not only web browser but all network applications including Outlook, Skype etc. all are dependent on Application layer protocols to function. There are dozens of application layer protocols that enable various functions at this layer. All these protocols collectively form the Application layer. These protocols form the basis for various network services like File transfer, Web surfing, Emails, Virtual terminals etc. File transfer is done with the help of FTP protocol. Web surfing is done with the help of HTTP or HTTPS For Emails, SMTP protocol is used, and For Virtual terminals, TELNET is used. So, Application Layer provides services for network applications with the help of protocols to perform users' activities (Turay, 2021).

Presentation layer receives data from Application Layer. This data is in the form of characters and numbers. Presentation Layer convert these characters and numbers to machine-understandable binary format Eg. Conversion of ASCII to EBCDIC code. This function of Presentation layer is called Translation. Before the data is transmitted, Presentation Layer reduces the number of bits that are used to represent the original data. This bit reduction process is called Data Compression and it can be Lossy or Lossless. Data Compression reduces the amount of space used to store the original file. As the size of file is reduced, it can be received at destination in very less time, i.e. data transmission can be done faster. Thus, data compression is very helpful in real-time video or audio streaming. To maintain the integrity of data, before transmission, data is encrypted. Encryption enhances security of the sensitive data. At senders side, data is encrypted and on the receiver's side, data is decrypted. SSL protocol (secure sockets layer) is used in Presentation Layer for Encryption and Decryption. So, Presentation Layer performs 3 basic function: Translation, Compression, and Encryption / Decryption (Froehlich et al., 2021).

Practical Case Study for Send/Receive Data Using Backend and Frontend With IOS Layers

Let's say the user wants to upload a video from a site www.example.com. When the user opens the Google Chrome browser and writes the name of the site that wants to download it www.example.com and press the send button note that the browser automatically wrote the word https: // before the site name www.example.com

This is the protocol responsible for the browsing process safely and the port number that works by 443 and this protocol is determined to do the process that the user wants by the application layer based on what the user wants to do and after receiving the application layer data to the presentation layer, which in turn performs four operations on this data

1: Encrypts the data sent by the user
2: Converts the data to the ASCII code
3: Compresses the data
4: Performs graphics processing

When sending a user request to show the site it receives and when the request returns and the appearance of the site www.example.com on the browser

Thus, the browser has dealt with the Front End site, which consists of the interfaces of the site resumed that the user sees images and words and expressions and everything the user can see and when the user wants to download a video file from the site, then the application layer to activate The protocol responsible for uploading and uploading FTB files is also sent to the application's presentation layer which in turn is based

1: Encrypts the data sent by the user
2: Converts the data to the ASCII code
3: Compresses the data
4: Performs graphics processing

Then send the request after the passage of the rest of the layers by the user then the application receives the server behind the site and this is called the backend, which is the side of the server-side and inside it all the databases of the site stored in it from the videos and passwords of users and others and the developer of the backend program This data is available in several languages including C # AND ORACLE AND SQL SERVER AND PHP AND JAVA. Upon receipt of the server, the user sends the video to him. This data-loaded video returns traffic through all IOS layers, including the presentation layer.

1: Decrypt the data
2: Decrypt ASCII code
3: Decompress the data
4: Reprocessing graphics

And then send the data to the application layer, which in turn shows what the user requested

Figure 31. Data traffic from the user to the database

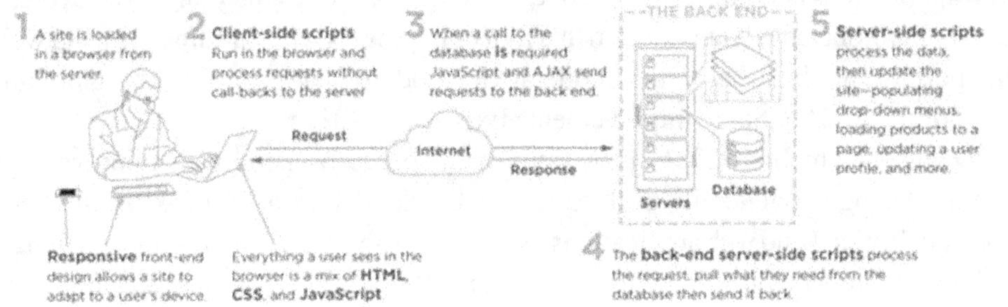

Tools Used for Designing the Website

A framework, or software stack, is a software application development platform. This provides a basis on which developers of software can create programs for a particular platform. For example, a framework may include predefined classes and functions for processing data, managing hardware devices, and communicating with system software. It streamlines the process of development as programmers do not need to reinvent the wheel each time a new application is created. And it is also a framework is a group of libraries of optimized and field-tested code that provide building blocks you can use to construct a website, The front-end frameworks or backend (Salas-Zárate et al., 2015).

A framework is similar to an interface for application programming (API), although an API is included in a framework. A framework acts as a foundation for programming, as the name suggests, while an API provides access to the elements provided by the framework. A system may also include libraries of code, a compiler, and other programs used in the process of software development. There are countless frameworks that developers use to build backend web apps, but the top 3 backend frameworks that impressed developers around the world are mentioned below (Haris & Hasim, 2019).

Spring Boot (Zhang et al., 2021) is extraordinary compared to other backend structures that engineers depend on while performing backend web advancement errands. It is a Spring-based system that enables engineers to compose generation grade backend web applications in Java. Spring Boot is microservices-prepared, which enables engineers to divvy the heap for appropriate treatment of solicitations made by the clients crosswise over gadgets and stages.

Express (Yule & Cooper, 2003)is A free, open-source backend web improvement structure, Express or Express.js is the highest system of Node.js. It is non-prohibitive in nature and is generally utilized by engineers over the world to compose backend web applications and fabricate effective APIs. Discharged in the year 2010, Express is an unopinionated Node.js system that assists engineers with building backend web and portable applications with a moderate methodology. It is one of the significant parts of the MEAN Stack improvement (Express.js, 2017).

Django (Campbell, 2021) is A free, open-source backend web improvement system, Django is generally utilized for the quick advancement of APIs and very good quality backend web applications. Named after the prestigious guitarist, Django Reinhardt, this Python structure was made in 2003. The Django structure underpins the speedy advancement of backend web applications with next to no coding. The effectiveness of this Python web improvement structure is demonstrated by the way that it is utilized by the absolute busiest applications on the web.

Bootstrap (Tutorials Point, 2017) is a system that can be utilized to rapidly plan a site. Bootstrap contains pre-composed CSS and JavaScript documents in which numerous classes are pre-characterized. These classes are utilized to make typography, HTML structures, pages responsive.

Responsive site is viewed as useful for SEO on the grounds that this kind of site decreases both page load time and page load speed contingent upon the gadget. On the off chance that somebody utilizes 3G or 4G information, at that point their information is additionally less. Numerous classes are now composed, which can be utilized over and over, this makes the structuring and improvement process quicker. Bootstrap can be come down to three principle records:

- bootstrap.css - a CSS structure
- bootstrap.js - a JavaScript/jQuery structure
- glyphicons - a textual style (a symbol text style set)

Furthermore, Bootstrap requires jQuery to work. jQuery is an incredibly famous and broadly utilized JavaScript library, that both rearranges and adds cross program similarity to JavaScript. Most tools used above framework to design and develop website, such as:

1. Microsoft Visual Studio

Visual Studio Code is perhaps the best JavaScript ide for Windows, Mac, and Linux. In addition to the fact that it supports JavaScript, however, it additionally bolsters Node.js, TypeScript, and it accompanies an entire biological system of augmentations for different dialects including C++, C#, Python, PHP and so on.

It gives extraordinary sentence structure featuring and auto-complete with IntelliSense dependent on factor types, word definitions, and imported modules. It likewise enables you to troubleshoot code by propelling or appending to your running applications investigate with breakpoints, call stacks, and an intelligent comfort. You can easily coordinate a JavaScript UI library into Visual Studio Code. With everything taken into account, this IDE for JavaScript is unquestionably one to look at. A significant note it is allowed to utilize. To get the most profitability out of the Visual Studio Code, we suggest this fast 1-hour online course, which will control you through including custom hotkeys, making formats and standards to speed up, coordinating with GitHub to have the option to deal with Git stores without leaving VS Code (Nawaz et al., 2022).

Figure 32. Microsoft visual studio

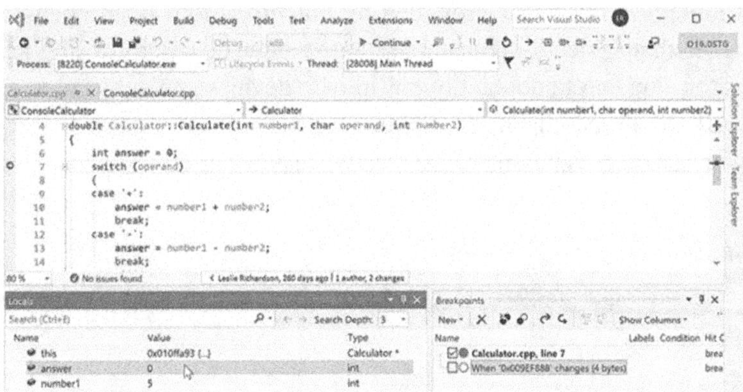

The various compatibility testing tools are useful for checking whether your web app functions on different devices and browsers, while its remarkable versatility makes it the best IDE for both beginners and seasoned developers.

Figure 33. Compatibility testing tools

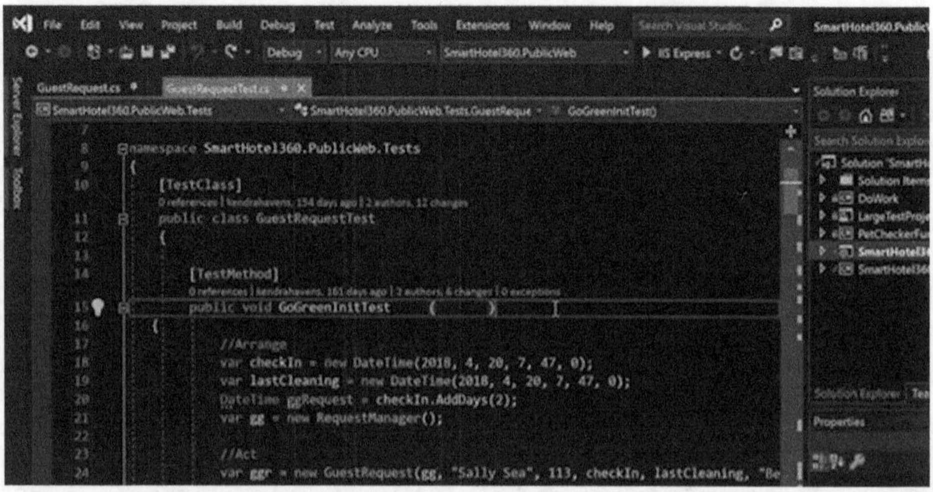

2. Atom

A cutting-edge supervisor propelled by GitHub, Atom is an open-source JavaScript IDE that is deeply adjustable. On account of a helpful implicit bundle administrator, you can undoubtedly put in new devices and subjects to extend its usefulness. Particle accompanies a few pre-introduced language structure subjects, has a perfect and clear view, and works crosswise over various working frameworks.

Figure 34. Atom

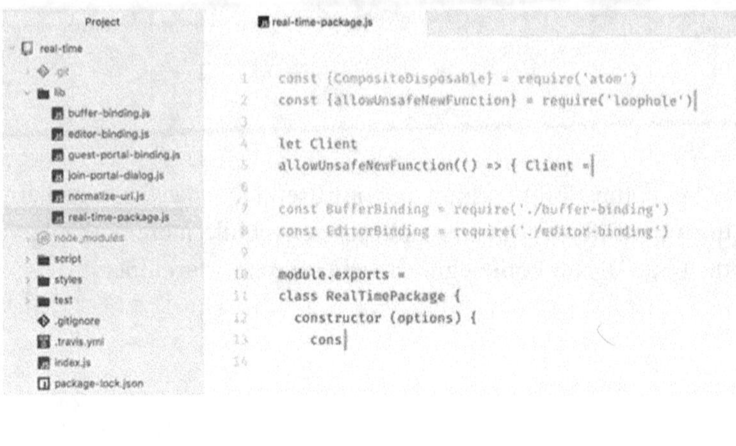

Atom by Github is JavaScript's best editor if you're looking for something that's easy to configure and use. It has an integrated package manager within this awesome tool to install new packages or start creating your own. Atom comes pre-installed in a variety of colors with four UIs and eight syntax themes. The rich and welcoming group also creates interesting templates that everyone can use to figure out what you're looking for.

Here are some of the best characteristics of the Atom:

- It works across different operating systems such as OS X, Windows, or Linux
- Find, preview, and replace text as you type in a file or across all your projects.

Easily browse and open a single file, a whole project, or multiple projects in one window.

3. Komodo

Another IDE that is extraordinary for JavaScript ventures is Komodo. It is a JavaScript IDE that has a streamlined interface and supports most significant dialects. It autocompletes code, highlights refactoring instruments, and performs reliably on various working frameworks.

Figure 35. Komodo

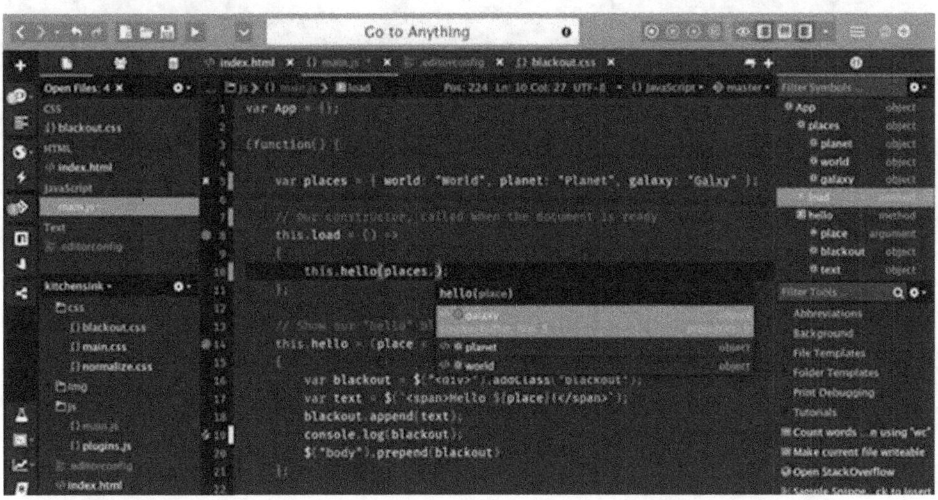

Notwithstanding standard highlights, Komodo additionally has some propelled usefulness, similar to single-step investigating and Syntax Checker. It additionally bolsters multi-client altering, making it helpful to work together on ventures inside your group (Eazley et al., 2017).

4. WebStorm

Created by JetBrains, WebStorm is a shut source programming item with a cutting-edge UI that fits directly into the best IDE for web improvement classification. It enables you to appreciate savvy code culmination, with helpful route and productive blunder discovery.

Figure 36. WebStorm

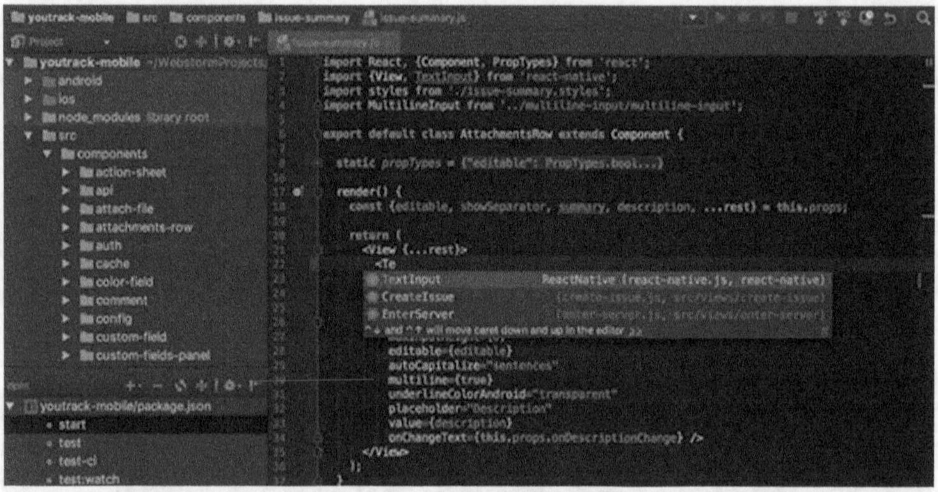

WebStorm has a powerful refactoring tool for style sheet languages and JavaScript and features an in-built debugger that will help you during web app development. WebStorm is also integrated with GitHub and Mercurial. All these features make it one of the best IDE for web development (Duarte et al., 2008).

5. IntelliJ IDEA

IntelliJ IDEA is a great web development IDE offering multiple plans. There is a free community version, but you should consider checking out the paid-for Ultimate Edition if you want to take advantage of all the Java Script development tools it has to offer. In reality, it might be worth your time. IntelliJ IDEA is a decent

CSS IDE, but it also supports a wide variety of programming languages, including AngularJS, CoffeeScript, HTML, JS, LESS, Node Javascript, PHP, Python, Ruby, Sass, and more.

The most important characteristics are:

- Extensive database editor and UML designer
- Supports multiple build systems
- Test runner UI
- Code coverage
- Git integration
- Deployment and debugging tools for most application servers
- Intelligent text editors for HTML, CSS, and Java
- Integrated version control
- AIR Mobile supports Android and iOS devices

Figure 37. IntelliJ IDEA

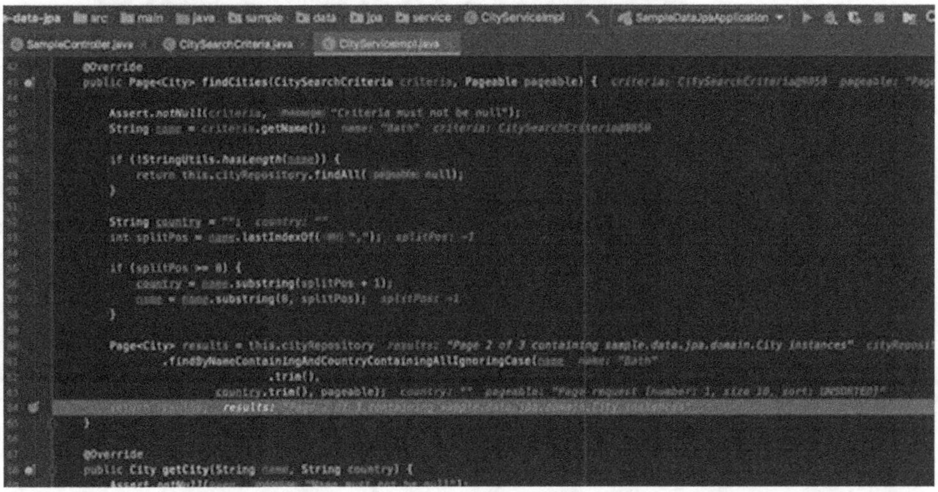

This phenomenal JavaScript IDE highlights a database supervisor, bolsters distinctive form frameworks, and gives organization and troubleshooting instruments. Also, there are a great deal of modules gave by JetBrains just as outsider designers that can spare you a ton of time (D'Arcy, 2009).

6. Sublime Text

The first on this rundown is a lightweight code manager considered Sublime Text that supports different dialects including HTML. This cross-stage programming has everything that the best IDE for web improvement ought to have

Figure 38. Sublime text

Sublime is a popular IDE and one of the best freely available JavaScript editors. It's running on Windows, Mac and Linux at the moment. It is quick and versatile and will do everything for website development that one should expect from the best IDE. Most users, even those who have been operating with Sublime for many years, do not know that it can be configured as a fully featured IDE, with features such as autocompletion of code, use of snippets and macros, etc. A 2-hour video course can enable beginners as well as experienced users to adapt Sublime to the best way for them to function (Ikhsan et al., 2020).

7. NetBeans

Figure 39. NetBeans

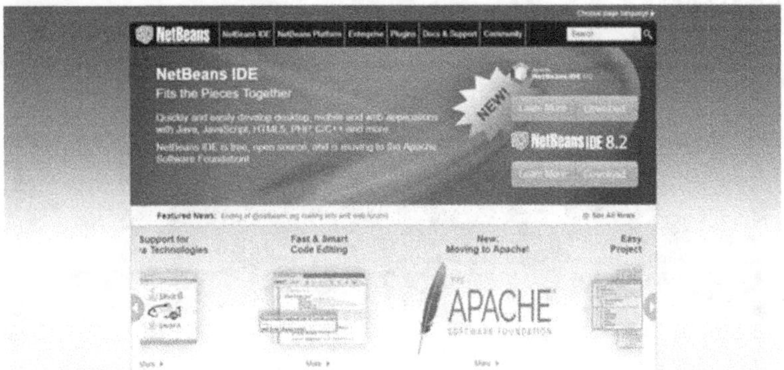

NetBeans is high on the rundown for the best web improvement IDE in light of the fact that it is anything but difficult to utilize and it lets you create cool work area, portable, and web applications in a matter of moments. It works similarly as great with JavaScript, HTML5, PHP, C/C++ and so forth. It is a free JavaScript IDE and an incredible HTML5 IDE for your everyday use.

This web advancement IDE accompanies cool code breaking down and altering devices perfect with the most recent Java 8 advances. This makes NetBeans 8.1 truly outstanding if not the best JavaScript editorial manager. It is an extraordinary AngularJS IDE just as a phenomenal apparatus for working with Node.js, Knockout.js and so forth. Notwithstanding all that, it is accessible in an assortment of dialects including English, Brazilian Portuguese, Japanese, Russian, and Simplified Chinese. As referenced before, NetBeans bolsters a wide scope of programming dialects, however in the event that you need the coolest web advancement IDE for your ventures, we prescribe downloading the HTML5/JavaScript manufactured establishment from the download page. As NetBeans IDE is a gigantic bundle, arranging nature, setting up the debugger, sending and executing an application could be precarious when you're simply beginning – this short NetBeans compressed lesson will assist you with getting started (NetBeans, 2015).

8. XCode

Among different instances of IDE, XCode merits a notice because of the wide scope of apparatuses made explicitly for applications running on Apple gadgets. It is completely incorporated with well-known systems, for example, Cocoa and

Cocoa Touch, enabling clients to include Passbook, Game Center, and different administrations with a solitary snap.

Figure 40. Xcode

Utilizing XCode, engineers can profit by the Apple LLVM compiler that adequately decides conceivable presentation issues, and Interface Builder that empowers fabricating a model without coding. Likewise, this IDE can consequently dissect and test code at any phase of advancement (Apple Inc, 2017).

9. Adobe Dreamweaver

Adobe Dreamweaver is a product program for planning website pages, basically an all the more completely highlighted HTML web and programming supervisor. The program gives a WYSIWYG (what you see is the thing that you get) interface to make and alter website pages. Dreamweaver bolsters numerous markup dialects, including HTML, XML, CSS, and JavaScript. Concerning human dialects, it underpins English, Spanish, French, German, Japanese, Chinese (both improved and customary), Italian, Russian, and some more. Dreamweaver was initially created and distributed by Macromedia in 1997. Adobe obtained Macromedia (which incorporated the rights to Dreamweaver) in 2005 and proceeded with the advancement of the program. The numerous highlights of Dreamweaver make it an adaptable web altering device, where it be for making mind boggling or very simples' locales (Firmansyah Agustian, 2017).

RESULTS

This chapter delved into the fundamental technologies underpinning modern websites, from server-side to client-side. We explored the crucial role of Domain Name Systems (DNS) in translating user-friendly domain names into numerical IP addresses, facilitating website access. The importance of DNS servers, specifically root servers, TLD servers, and authoritative nameservers, was highlighted, emphasizing how they work together to resolve domain names efficiently. Then examined various internet protocols, focusing on their role in enabling communication between clients and servers. Key protocols like HTTP, HTTPS, FTP, SMTP, and TCP/IP were discussed, emphasizing their specific functionalities and importance in secure data transfer and website interaction.

The chapter further explored the concept of web servers, emphasizing their role as the foundation for hosting and serving website content. Different types of servers like proxy servers, mail servers, application servers, and virtual servers were presented, highlighting their unique functions and contributions to the overall website ecosystem. Finally, the chapter delved into the world of website development, outlining the key differences between the front-end (client-side) and back-end (server-side) technologies. We explained how front-end developers focus on creating visually appealing and user-friendly interfaces using HTML, CSS, and JavaScript, while back-end developers handle the logic and functionality of the website, often using languages like PHP, Java, or Python.

The importance of a well-designed user interface (UI) and user experience (UX) was stressed, emphasizing how they contribute to a successful website. The concept of full-stack development was also discussed, highlighting the skills and knowledge required for developers to be proficient in both front-end and back-end development. Furthermore, we explored the OSI model, a standardized framework for understanding how data travels across networks. Each of the seven layers was briefly described, explaining how they work together to enable seamless communication between computers.

CONCLUSION

The chapter concluded with a practical case study demonstrating the process of registering a domain name and a practical example of data traffic flow from a user request through the front-end, back-end, and various OSI layers. This real-world application demonstrated the interconnected nature of these technologies in creating a functioning website.

In conclusion, this chapter provided a comprehensive overview of the fundamental technologies essential for understanding and building modern websites. It equips readers with a solid understanding of the concepts and tools necessary to create engaging and functional online experiences. By gaining insight into the interplay of DNS, protocols, server technologies, and website development methodologies, readers are better equipped to navigate the complex world of web development and design.

REFERENCES

Apple Inc. (2017). Xcode IDE. In *Apple Developer*. Author.

Aziz, B. (2020). Modeling and Analyzing an Industry 4.0 Communication Protocol. *IEEE Internet of Things Journal*, 7(10), 10120–10127. Advance online publication. 10.1109/JIOT.2020.2999325

Baiskar, Y. (2022). MERN: A Full-Stack Development. *International Journal for Research in Applied Science and Engineering Technology*, 10(1), 1029–1035. Advance online publication. 10.22214/ijraset.2022.39982

Barrett, K., Hunt, C. J., Lange, L., & Meyer, A. S. (2020). Conserved unique peptide patterns (CUPP) online platform: Peptide-based functional annotation of carbohydrate active enzymes. *Nucleic Acids Research*, 48(W1), W110–W115. Advance online publication. 10.1093/nar/gkaa37532406917

Bell, C. P. M., Rees, J. M., Naylor, T., Mayne, N. J., Jeffries, R. D., Mamajek, E. E., & Rowe, J. (2014). Pre-main-sequence isochrones - III. The cluster collaboration isochrone server. *Monthly Notices of the Royal Astronomical Society*, 445(4), 3496–3511. Advance online publication. 10.1093/mnras/stu1944

Brownlee, N., Claffy, K. C., & Nemeth, E. (2001). DNS measurements at a root server. *Conference Record / IEEE Global Telecommunications Conference, 3*. 10.1109/GLOCOM.2001.965864

Campbell, S. (2021). *Flask vs Django: What's the Difference Between Flask & Django?* Guru99.

Christanto, F. W., & Sani Suprayogi, M. (2018). Enhancement Network Monitoring System Functionality by Implementing an Android-based Notification System to Monitor Virtual Servers on Cloud Computing Network. *IJAIT*, 2(1), 7. Advance online publication. 10.25124/ijait.v2i01.1002

Cloudflare. (2020). *What Is DNS? | How DNS Works | Cloudflare*. Cloudflare Inc.

D'Arcy, H. (2009). *IntelliJ IDEA*. RefCardz.

Duarte, P., Costa, C. J., & Costa, P. (2008). *Webstorm*. 10.1145/1456536.1456592

Dziallas, M. (2020). How to evaluate innovative ideas and concepts at the front-end?: A front-end perspective of the automotive innovation process. *Journal of Business Research*, 110, 502–518. Advance online publication. 10.1016/j.jbusres.2018.05.008

Eazley, D., Jones, B. K., Martelli, A., Ravenscroft, A., Holden, S., Schlusser, T., Reitz, K., Stephenson, B., & Dwyer, G. (2017). The Hitchhiker's Guide to Python – Best Practices for Development. In *Packt Publishing* (Vol. 44, Issue 8).

Express.js. (2017). *Express - Node.js web application framework*. Expressjs.Com.

Firmansyah, A. S., & Riadi, I. (2014). Analisis dan perancangan proxy server menggunakan virtual machine. *Sarjana Teknik Informatika*.

Firmansyah Agustian. (2017). Adobe Dreamweaver. *Jurnal Desain Web, 1*.

Froehlich, A., Rosencrance, L., & Gattine, K. (2021). What is the OSI model? The 7 layers of OSI explained. *Tech Target*.

Garais, G. E., & Enaceanu, A.-S. (2016). Open Source Servers and Website Platforms Security. *Journal of Information Systems & Operations Management*.

Haris, N. A., & Hasim, N. (2019). PHP frameworks usability in web application development. *International Journal of Recent Technology and Engineering, 8*(3). 10.35940/ijrte.C1020.1083S19

Ikhsan, A., Ma, S., Perdananto, A., Studi Magister Informatika, P., Eresha, S., Selatan, T., & Raya Puspitek No, J. (2020). Aplikasi Pemesanan Tiket Seminar Secara Online Berbasis Web Di Universitas Pamulang. *Jurnal Teknologi Informasi ESIT*.

Kumar, V., & Roy, O. P. (2021). Security and challenges in voice over internet protocols: A survey. *IOP Conference Series. Materials Science and Engineering*, 1020(1), 012020. Advance online publication. 10.1088/1757-899X/1020/1/012020

Mahler, T. (2014). A gTLD right? Conceptual challenges in the expanding internet domain namespace. *International Journal of Law and Information Technology*, 22(1), 27–48. Advance online publication. 10.1093/ijlit/eat018

McGillivray, K. (2014). Give it away now? Renewal of the IANA functions contract and its role in internet governance. *International Journal of Law and Information Technology*, 22(1), 3–26. Advance online publication. 10.1093/ijlit/eat017

Nawaz, A., Arora, A. S., Ali, W., Saxena, N., Khan, M. S., Yun, C. M., & Lee, M. (2022). Intelligent Human-Machine Interface: An Agile Operation and Decision Support for an ANAMMOX SBR System at a Pilot-Scale Wastewater Treatment Plant. *IEEE Transactions on Industrial Informatics*, 18(9), 6224–6232. Advance online publication. 10.1109/TII.2022.3153468

NetBeans. (2015). NetBeans IDE. *Features, Jsr 299*.

Noble, E. (2023). Deployment. *Pro T-SQL*, 2022, 321–345. 10.1007/978-1-4842-9256-3_12

Oliveira, L., Rodrigues, J. J. P. C., Kozlov, S. A., Rabêlo, R. A. L., & de Albuquerque, V. H. C. (2019). MAC layer protocols for internet of things: A survey. In *Future Internet* (Vol. 11, Issue 1). 10.3390/fi11010016

Petrikoglou, A., & Kaskalis, T. H. (2019). Full stack web development teaching: Current status and a new proposal. *WEBIST 2019 - Proceedings of the 15th International Conference on Web Information Systems and Technologies*. 10.5220/0008066202180225

Raja, J. Z., Chakkol, M., Johnson, M., & Beltagui, A. (2018). Organizing for servitization: Examining front- and back-end design configurations. *International Journal of Operations & Production Management*, 38(1), 249–271. Advance online publication. 10.1108/IJOPM-03-2016-0139

Salas-Zárate, M. D. P., Alor-Hernández, G., Valencia-García, R., Rodríguez-Mazahua, L., Rodríguez-González, A., & López Cuadrado, J. L. (2015). Analyzing best practices on Web development frameworks: The lift approach. In *Science of Computer Programming* (Vol. 102). 10.1016/j.scico.2014.12.004

Syarifudin, A. (2020). *Konfigurasi DHCP Server*. Fakultas Komputer.

Turay, B. (2021). Analysis of Seven Layered Architecture of Osi Model. SSRN *Electronic Journal*. 10.2139/ssrn.3815237

Tutorials Point. (2017). *Bootstrap responsive web development*. Tutorials Point.

Weinberg, J. (2000). ICANN and the problem of legitimacy. *Duke Law Journal*, 50(1), 187. Advance online publication. 10.2307/1373114

Wilkowski, M. (2021). Was this the real Web? Quantitative overview of the Polish... ccTLD Internet Archive data (1996–2001). *Archeion*, 122, 44–68. Advance online publication. 10.4467/26581264ARC.21.015.14495

Williams, T., Vo, H., Samset, K., & Edkins, A. (2019). The front-end of projects: A systematic literature review and structuring. *Production Planning and Control*, 30(14), 1137–1169. Advance online publication. 10.1080/09537287.2019.1594429

Wulandari, H., Suherman, S., & Razali, R. (2022). Pelatihan administrasi sistem jaringan berbasis windows pada guru dan siswa (Studi kasus: Sekolah smk madani marendal I). *RESWARA. Jurnal Pengabdian Kepada Masyarakat*, 3(1), 185–190. Advance online publication. 10.46576/rjpkm.v3i1.1637

Xian, F., Srisa-an, W., & Jiang, H. (2008). Garbage collection: Java application servers' Achilles heel. *Science of Computer Programming*, 70(2–3), 89–110. Advance online publication. 10.1016/j.scico.2007.07.008

Xu, H., Zhang, Z., Han, B., & Yan, J. (2021). Identifying Influential SLD Authoritative Name Servers on the Internet. *Frontiers in Physics (Lausanne)*, 9, 768205. Advance online publication. 10.3389/fphy.2021.768205

Yule, P., & Cooper, R. P. (2003). Express: A Web-based technology to support human and computational experimentation. *Behavior Research Methods, Instruments, & Computers*, 35(4), 605–613. Advance online publication. 10.3758/BF0319554014748505

Zammetti, F. (2020). Modern Full-Stack Development. In *Modern Full-Stack Development*. 10.1007/978-1-4842-5738-8

Zhang, F., Sun, G., Zheng, B., & Dong, L. (2021). Design and implementation of energy management system based on spring boot framework. *Information (Basel)*, 12(11), 457. Advance online publication. 10.3390/info12110457

Zimmermann, H. (1980). OSI Reference Model-The ISO Model of Architecture for Open Systems Interconnection. *IRE Transactions on Communications Systems*, 28(4), 425–432. Advance online publication. 10.1109/TCOM.1980.1094702

Zou, L., Wang, Z., Hu, J., Liu, Y., & Liu, X. (2021). Communication-protocol-based analysis and synthesis of networked systems: Progress, prospects and challenges. *International Journal of Systems Science*, 52(14), 3013–3034. Advance online publication. 10.1080/00207721.2021.1917721

Section 2
Analysis, Architecting, and Designing Software

Comprises the requirements specification, which is the first stage of developing a software system. The requirements specification can be considered as the agreement between developer and your customer. In the requirement specification, this section describe your software system that is intended to solve the customer's problem. It basically includes the functionality of the system (i.e., what the software does and does not), the system's interaction with the users, quality requirements, and design constraints, as a result reader will be able to Justify the tools and techniques chosen to realis a custom built website for design phase. This section clearly describe the components of your software system and their relationships, it also describe all necessary information about these components so that programmers can use your design to implement the software system.

Chapter 2
Software Process Model for Website

ABSTRACT

This chapter introduces readers to the underpinning services required for software engineering for modern web applications, how to analyse, developing requirements, and modelling with classes using UML. First, the authors introduce software and software engineering to help new designers to model interactions and behaviour using UML. Then they show how to justify the difference between software and software engineering, and analysis and developing requirements for modelling using UML, which process model to use in SW process, and why.

BACKGROUND

Software engineering (Khan et al., 2021) is the discipline concerned with the application of theory, knowledge, and practice for effectively and efficiently building software systems that satisfy the requirements of users and customers. Software engineering is applicable to small, medium, and large-scale systems. It encompasses all phases of the life cycle of a software system. Software engineering employs engineering methods, processes, techniques, and measurements. It benefits from the use of tools for managing software development; analyzing and modeling software artifacts; assessing and controlling quality; and ensuring a disciplined, controlled approach to software evolution and reuse. Software development, which often involves a team of developers, requires choosing the suitable tools, methods, and approaches that are most applicable to a given development environment. Software engineering is an engineering discipline that is concerned with all aspects of software production from the early stages of system specification through to maintaining the system after it has gone into use (Brereton et al., 2007), as software engineering is important to differentiate and be clear between program and software, as program a

DOI: 10.4018/978-1-6684-4844-1.ch002

set of instructions written in a particular programming language for a specific purpose, while software a combination of the program(s), documentation (development documents) and operating procedure documents (provided to customers at the time of release), the software should deliver the required functionality and performance to the user and should be maintainable, dependable and usable, as described below:

- Maintainability, Software must be able to evolve to meet changing needs with minimal effort and time
- Dependability, Software must be trustworthy
- Efficiency, Software should not make wasteful use of system resources
- Acceptability and Usability, Software must be acceptable and usable by the users for the purpose it was designed for.

In software engineering, you should work on the engineering discipline by using appropriate theories and methods to solve problems bearing in mind organizational and financial constraints using all aspects of software production not just the technical process of development (Hilburn et al., 1999). Also, project management and the development of tools, methods, etc. to support software production, Figure 2 shows the software engineering body of knowledge.

Figure 1. Software engineering body of knowledge (Hilburn et al., 1999)

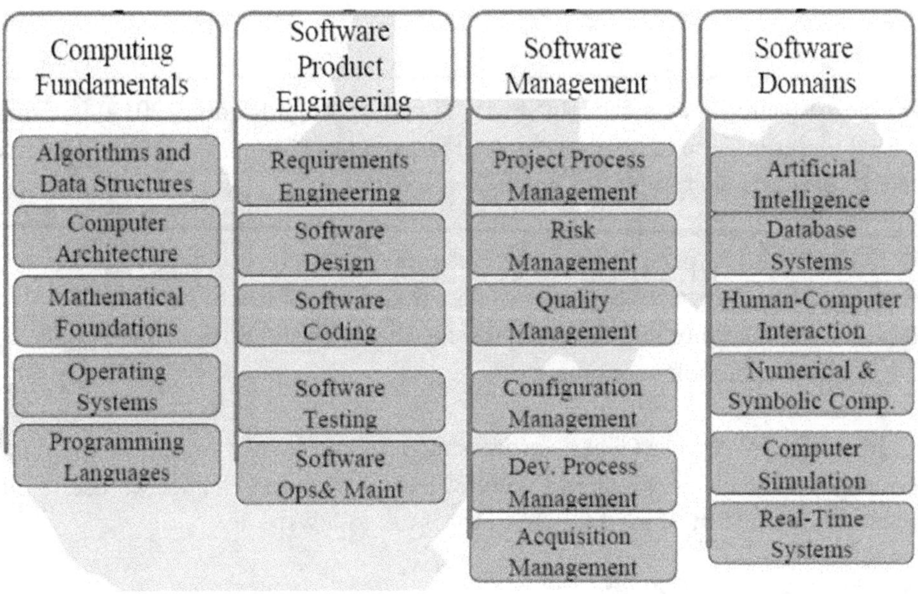

Costs of software are a very important topic for the software analysis phase, as it varies depending on the type of system being developed and the requirements of system attributes, for example for high system performance and reliability costs can be high (Heemstra, 1992). Roughly 60% of costs are development costs, 40% are testing costs. For custom software, evolution costs often exceed development costs. Distribution of costs depends on the development model that is used as shown in Table 1.

Table 1. Software cost

System Development	
System Requirements	2
Hardware Requirements	8
Software Requirements	10
Software Design	12
Coding	13
Unit Test	24
Integration Test	13
Documentation	6
System Test	12
TOTAL	**100**

The software Process Model (SPM) (Tiwari & Singh Rathore, 2019) is a simplified representation of a software process, presented from a specific perspective and we have many examples of process perspectives:

- The workflow perspective represents inputs, outputs, and dependencies
- Data-flow perspective represents data transformation activities
- The role/action perspective represents the roles/activities of the people involved in the software process

In system SW we use Generic process models, and each of the models used has its details and uses as shown in Figure 2, later in the next chapter we use in our practical case, the most used Generic process as follows:

- **Waterfall** (Casteren, 2017)
- **Evolutionary development (commonly known as agile)** (Casteren, 2017)
- **Formal transformation** (Dickerson et al., 2018)

- **Reuse-oriented: Integration from reusable components** (Chu et al., 2000)
- **SEMLHI** (*SEMLHI*, n.d.)
- **DevOps** (Leite et al., 2019)

Figure 2. DevOps model (How To Become A DevOps Engineer In 2022: A Comprehensive Guide, n.d.)

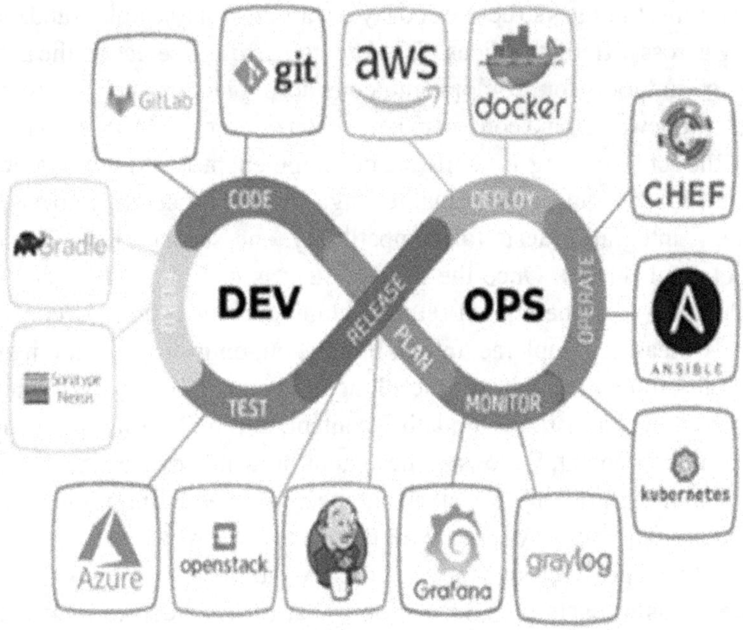

Software Requirements Specification (SRS), is an official statement of what the system developers should implement, It should include both the user requirements for a system and a detailed specification of the system requirements, Agile methods do not use official SRS, instead, they use User Stories, for Critical and Outsourcing Projects needs detailed SRS document.

ANALYSIS AND DEVELOPING REQUIREMENTS

A software process is a set of related activities that leads to the production of a software product, while user requirements should be simple natural language who don't have a technical background, and it can have some intuitive diagrams such as use case diagrams. In general, the system requirements should be, Correct, Consistent, and Complete.

Developing a website involves a series of steps that demand careful planning and execution. Firstly, it is crucial to thoroughly analyze the project's objectives and target audience, as well as understand the users' requirements and desired features. This initial phase also includes conducting feasibility studies and risk assessments to ensure project viability. Subsequently, the focus shifts to designing the user interface (UI) and user experience (UX) elements. Wireframes and mockups are created to outline the site's structure and layout, aiming for a seamless and responsive experience across different devices. The next step involves selecting the appropriate technologies and tools for development, considering factors such as programming languages, frameworks, and content management systems (CMS). With the groundwork laid, the actual coding and software development process begins, emphasizing clean code practices and continuous testing. Testing procedures cover unit tests for accuracy, integration tests for compatibility, and security tests for protection against potential threats. Once the development is complete, the website undergoes deployment and operational setup, including hosting, domain configuration, and security measures implementation. Finally, ongoing support and maintenance are essential for ensuring the website's longevity and effectiveness. This involves providing technical assistance, updating content, and monitoring performance for continuous improvement. Following these steps diligently can lead to the successful development of high-quality websites that meet user needs and project objectives.

The requirements for a system are the descriptions of what the system should do, the services that it provides, and the constraints on its operation. These requirements reflect the needs of customers for a system that serves a certain purpose such as controlling a device, placing an order, or finding information. The detailed requirements serve as a contract between the development company and the customer.

Functional Requirements: These are statements of services the system should provide, how the system should react to particular inputs, and how the system should behave in particular situations (Benfell, 2021). Figure 3 shows the following points which represent the functional requirement:

1. Students can access exams online using computer or smartphone devices.
2. English teachers will get students' grades automatically after the last student submits his attempts on the time end for the exam.
3. The teacher will add marks for each question in the exam and total exam mark will depends on summation of total question marks.
4. Responses by the candidates will be checked automatically.
5. time consuming, Staff cost and paper work will reduce.

6. Teacher can create different common type of question include multiple options, multiple answers, true false, fill in blank.
7. Student can get exam result automatically after submitting and finishing exam.

Figure 3. User requirements with functional requirements that will use by users

Non-Functional Requirements

In addition to the functional requirements that dictate the specific features and capabilities of the website, attention must also be given to non-functional requirements. These requirements define the qualities and characteristics of the system, such as performance, reliability, security, and scalability. For instance, the website must be designed to handle a certain number of concurrent users without compromising speed or responsiveness, ensuring a positive user experience. Security measures should be implemented to safeguard user data and protect against potential cyber threats. Reliability is crucial, necessitating regular backups and failover mechanisms to minimize downtime. Scalability considerations ensure that the website can accommodate growth and increased usage over time. By addressing non-functional requirements alongside functional ones, the development team can deliver a website that not only meets user needs but also excels in terms of performance, security, and reliability.

These are constraints on the services or functions offered by the system, and we can classify to the following type (Shankar et al., 2020):

Non-Functional Requirements Example

Let's consider we work on project for "Smart Company", an e-commerce platform specializing in electronics. Based on the aforementioned text, here's how we can derive non-functional requirements:

- Performance characteristics, the Smart Company website must be able to handle a high volume of concurrent users, especially during peak times such as sales events or promotions. For example, the website should be capable of processing a minimum of 1000 simultaneous transactions without experiencing significant slowdowns or downtime.
- Resources and Management Issues, given the importance of uninterrupted service, the Smart Company website must have a robust backup and disaster recovery system in place. For instance, daily automated backups should be conducted, and a failover mechanism should be implemented to ensure minimal downtime in the event of server failures.
- Security Issues, as an e-commerce platform dealing with sensitive customer information, the Smart Company website must prioritize security. For example, the website should use HTTPS encryption to protect user data during transmission, implement multi-factor authentication for administrative access, and regularly undergo security audits and vulnerability assessments.
- Quality issues: with the potential for rapid growth in customer base and transaction volume, the Smart Company website must be designed with scalability in mind. For instance, the architecture should be modular and scalable horizontally to accommodate increased traffic without requiring significant infrastructure changes.

By incorporating these non-functional requirements into the development process, Smart Company can ensure that its website not only meets the functional needs of its users but also delivers a seamless, secure, and reliable experience, even during periods of high demand.

Modelling With Classes Using UML

A UML use case diagram is the primary form of system/software requirements for a new software program underdeveloped. Use cases specify the expected behavior (what), and not the exact method of making it happen (how). Use cases once specified can be denoted both textual and visual representation (i.e. use case diagram). Main concept of use case modeling is that it helps us design a system from the end user's perspective. It is an effective technique for communicating system behavior in the user's terms by specifying all externally visible system behavior. Use case diagrams are typically developed in the early stage of development and people often apply use case modeling for the following purposes (Dickerson et al., 2018):

- Specify the context of a system
- Capture the requirements of a system
- Validate a systems architecture
- Drive implementation and generate test cases
- Developed by analysts together with domain experts

Use cases represent only the functional requirements of a system, use case modeling is often associated with UML, although it has been introduced before UML existed. Its brief history is as follow:

- In 1986, Ivar Jacobson first formulated textual and visual modeling techniques for specifying use cases.
- In 1992 his co-authored book Object-Oriented Software Engineering - A Use Case Driven Approach helped to popularize the technique for capturing functional requirements, especially in software development.

System modeling is the process of developing abstract models of a system, with each model presenting a different view or perspective of that system, figure 4 illustrate the main UML Models which used.

Figure 4. Main UML models which used

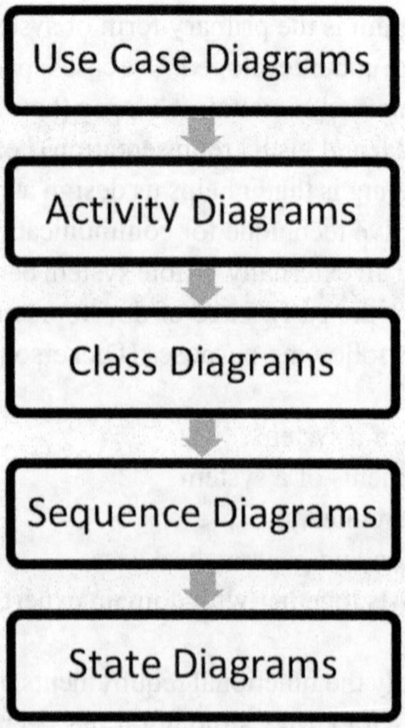

Use case modeling mostly used to model interactions between a system and external actors (users or other systems), A standard form of use case diagram is defined in the UML as shown in the Use Case Diagram in Figure 5.

Figure 5. Case diagram cellular phone example

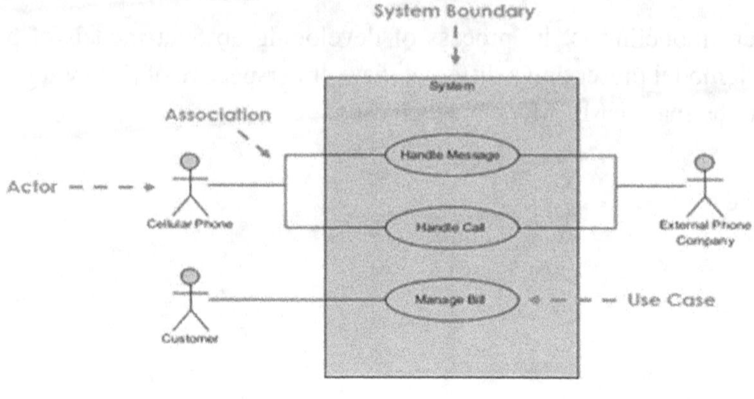

Actor: Actor plays a role in the business, someone interacts with use case (system function), normally Named by noun, similar to the concept of user, but a user can play different roles, for example, a prof. can be instructor and also researcher which plays two roles with two systems, the actor has a responsibility toward the system (inputs), and Actor has expectations from the system (outputs), Alternative notations as shown in Figure 6:

Figure 6. Actors alternative notations

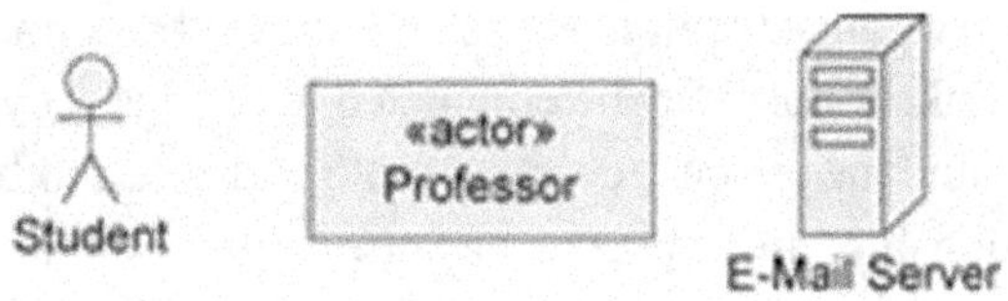

In software engineering it easiest to start the requirements elicitation process by identifying the actors. Below the questions can help you identify the actors of your system:

- Who uses the system?
- Who installs the system?
- Who starts up the system?
- Who maintains the system?
- Who shuts down the system?
- What other systems use this system?
- Who gets information from this system?
- Who provides information to the system?
- Does anything happen automatically at a present time?

Use Case

The use case diagram does not cover the internal structure and the actual implementation of a use case, it describes functionality expected from the system under development, it Provides tangible benefit for one or more actors that communicate with this use case, it derived from collected customer wishes. Set of all use cases describes the functionality that a system shall provide.

Each Actor must be linked to a use case, while some use cases may not be linked to actors. Named by verb + Noun (or Noun Phrase), i.e. Do something. A use case is triggered either by invocation of an actor or by a trigger event, in short, a trigger. Alternative notations as shown in Figure 7:

Figure 7. Use case notation

Communication Link

The participation of an actor in a use case is shown by connecting an actor to a use case by a solid link. Actors may be connected to use cases by associations, indicating that the actor and the use case communicate with one another using messages.

The extend relationships are important because they show optional functionality or system behavior. The <<extend>> relationship is used to include optional behavior from an extending use case in an extended use case as shown in example in Figure 8.

Figure 8. Communication link notation

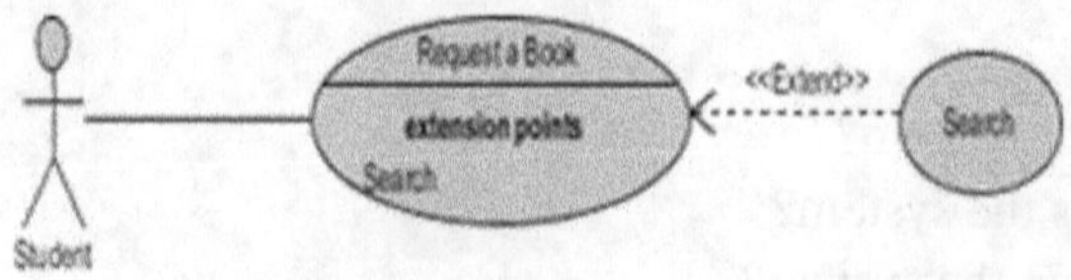

The include relationship is depicted with a directed arrow having a dotted line. The tip of arrowhead points to the child use case and the parent use case connected at the base of the arrow. The stereotype "<<include>>" identifies the relationship as an include relationship as shown in Figure 9.

Figure 9. <<include>> notation

The include relationship adds additional functionality not specified in the base use case. The <<Include>> relationship is used to include common behavior from an included use case into a base use case to support the reuse of common behavior as shown in Figure 10.

Figure 10. Use case include example

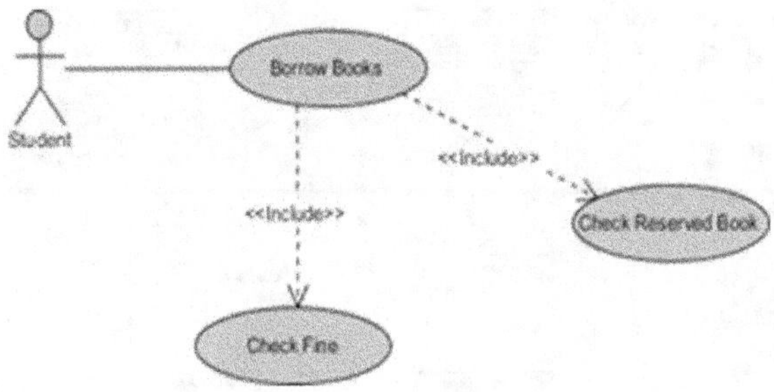

Activity diagrams helps to represent Workflows and business processes, it model the behaviors of the system, and show the dependencies and coordination between activities within a system, the activity flow should not get "stuck", it can be used during the requirements elicitation process to help identify how use cases interact to achieve business processes and to identifying use cases of a system for the operations involved in the realization of a use case, Finally, they can be attached to any model element to model its dynamic behaviour.

A Class diagram gives an overview of a system by showing its classes and the relationships among them, it static as they display what interacts but not what happens when they do interact.

next figure shows the class diagram models for a customer order from a retail catalog. The central class is the Order. Associated with it are the Customer making the purchase and the Payment. A Payment is one of four kinds: Cash, Check, Credit or Wire Transfer. The order contains OrderDetails (line items), each with its associated Item as shown in Figure 11 and Figure 12.

Figure 11. Sales order system: use case example

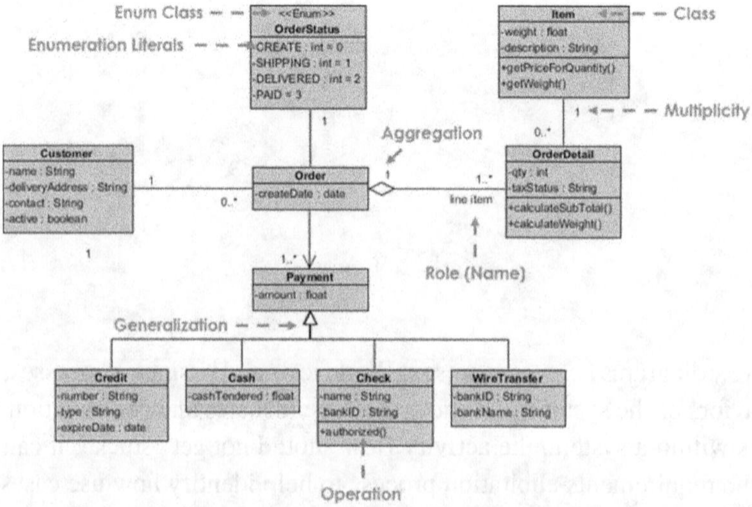

Modelling Interactions and Behaviour using UML – 1 for the use case is a fundamental concept of many object-oriented development methods. While Modelling Interactions and Behaviour using UML – 2 used in Domain names are the name of the site that is built by the designer, and this name is the only one of the names of all domain names in the world is the identity of the site.

Figure 12. Use case diagram: Vehicle sales systems example

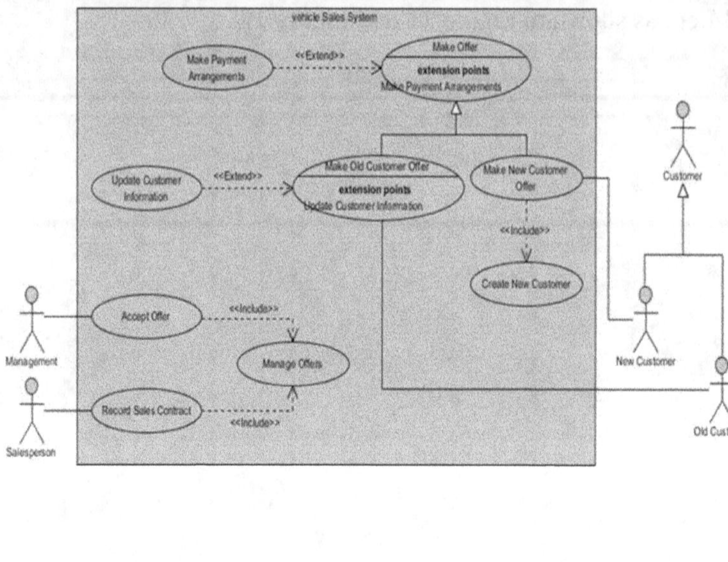

FOCUSSING ON USERS AND THEIR TASKS

A software architecture is a description of how a software system is organized, describe the architecture, explain when it may be used and describe its advantages and disadvantages. Software Architecture affects, Performance, Robustness, Disreputability, And Maintainability. Advantages of documenting software architecture for stakeholder communications it focuses of discussion, for system analysis allow him to view early stage of development, and for large-scale reuse, usually software products with similar requirements can be designed using similar software architecture.

Architecture Design is the critical link between design and requirements engineering, it identifies the main structural components in a system and the relationships between them. The output of the architectural design process is an architectural model that describes how the system is organized as a set of communicating components. Architecture design is a creative process and depends on:

- Type of system being developed
- Technology being used
- Experience and background of system architect/team lead

The Model-View-Controller (MVC) Pattern

Patterns are a means of representing, sharing and reusing knowledge, an architectural pattern is a stylized description of good design practice, which has been tried and tested in different environments, patterns should include information about when they are and when the are not useful, patterns may be represented using tabular and graphical descriptions (Cortez & Vazhenin, 2015).

MVC used when there are multiple ways to view and interact with data. Also used when the future requirements for interaction and presentation of data are unknown. The system is structured into three logical components that interact with each other. The Model component manages the system data and associated operations on that data. The View component defines and manages how the data is presented to the user. The Controller component manages user interaction (e.g., key presses, mouse clicks, etc.) and passes these interactions to the View and the Model. Next figure shows the architecture of a web-based application system organized using the MVC pattern (Paolone et al., 2020).

Figure 13. Web application architecture using the MVC pattern

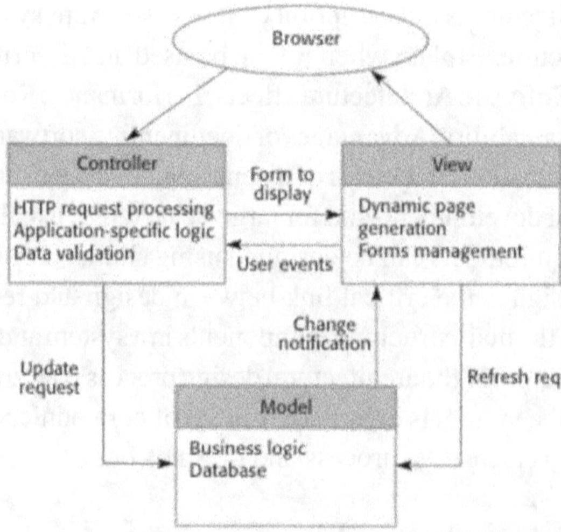

Layered Architecture

It used to model the interfacing of sub-systems by organises the system into a set of each of which provide a set of services, it supports the incremental development of sub-systems in different layers. When a layer interface changes, only the adjacent layer is affected.

Organizes the system into layers with related functionality associated with each layer, each layer provides services to the layer above it so the lowest-level layers represent core services that are likely to be used throughout the system (Martínez et al., 2019).

Used when building new facilities on top of existing systems; when the development is spread across several teams with each team responsibility for a layer of functionality; when there is a requirement for multi-level security. A layered model of Mobile App, as shown in Figure 14.

Figure 14. Application architecture for a mobile App: iOS, Android, Hybrid (Moreb, 2022)

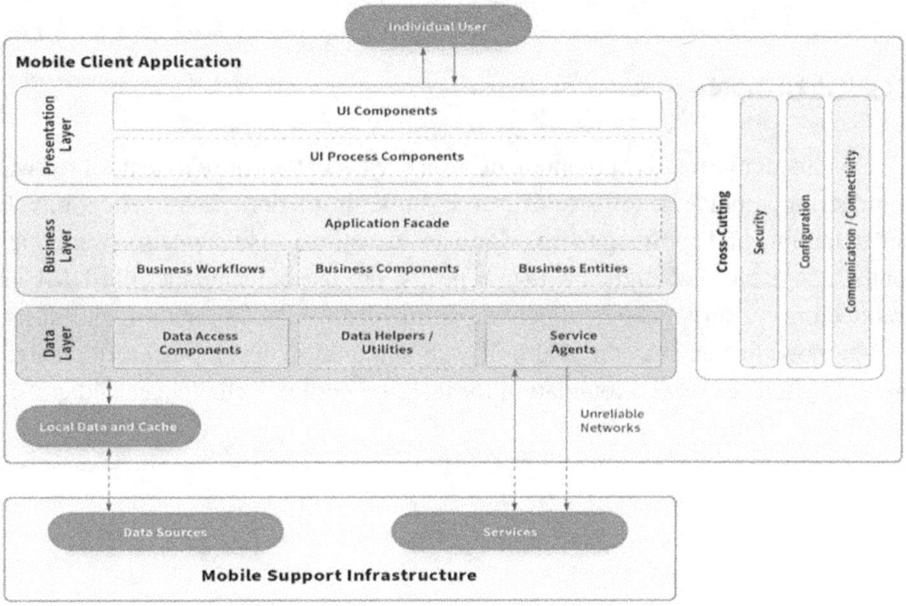

Results and Conclusion

This chapter delved into the foundational concepts of software engineering, focusing on its application in the development of modern web applications. We explored the key elements that comprise the software engineering process, starting with understanding the distinction between "software" and "software engineering."

Software engineering goes beyond simply writing code; it encompasses a structured approach that ensures quality, maintainability, and efficiency throughout the software lifecycle. This process begins with a crucial step: analyzing and developing requirements. This involves meticulously gathering, documenting, and refining user needs into functional and non-functional requirements, forming the bedrock upon which the system's design is built. UML plays a pivotal role in visualizing and communicating software design, enabling effective collaboration between developers and stakeholders. Use Case Diagrams, powerful tools for capturing user interactions with a system, offer a clear representation of intended functionality by identifying actors, use cases, and their relationships. Activity Diagrams provide a visual representation of workflows and business processes, helping to understand the flow of activities and dependencies within a system. Lastly, architecture design is essential for ensuring a well-structured and maintainable software system. Architectural patterns like MVC and layered architectures provide established frameworks for organizing system components, promoting robust and adaptable software.

CONCLUSION

This chapter provided a comprehensive overview of the core elements of software engineering, emphasizing its relevance in building modern web applications. By understanding and applying these principles, developers can create software that is robust, adaptable, and aligned with user needs. The use of techniques like UML and architectural patterns fosters effective communication, collaboration, and design. As we move further into the realm of web application development, the insights presented here serve as a foundation for tackling complex challenges and crafting innovative solutions.

REFERENCES

Benfell, A. (2021). Modeling functional requirements using tacit knowledge: A design science research methodology informed approach. *Requirements Engineering*, 26(1), 25–42. Advance online publication. 10.1007/s00766-020-00330-4

Brereton, P., Kitchenham, B. A., Budgen, D., Turner, M., & Khalil, M. (2007). Lessons from applying the systematic literature review process within the software engineering domain. *Journal of Systems and Software*, 80(4), 571–583. Advance online publication. 10.1016/j.jss.2006.07.009

Casteren, W. van. (2017). The Waterfall Model And The Agile Methodologies : A Comparison By Project Characteristics-Short The Waterfall Model and Agile Methodologies. *Academic Competences in the Bachelor*.

Chu, W. C., Lu, C. W., Yang, H., & He, X. (2000). A formal approach for component retrieval and integration analysis. *Journal of Software Maintenance: Research and Practice*, 12(6), 325–342. Advance online publication. 10.1002/1096-908X(200011/12)12:6<325::AID-SMR217>3.0.CO;2-#

Cortez, R., & Vazhenin, A. (2015). Virtual model-view-controller design pattern: Extended MVC for service-oriented architecture. *IEEJ Transactions on Electrical and Electronic Engineering*, 10(4), 411–422. Advance online publication. 10.1002/tee.22101

Dickerson, C. E., Roslan, R., & Ji, S. (2018). A Formal Transformation Method for Automated Fault Tree Generation from a UML Activity Model. *IEEE Transactions on Reliability*, 67(3), 1219–1236. Advance online publication. 10.1109/TR.2018.2849013

Heemstra, F. J. (1992). Software cost estimation. *Information and Software Technology*, 34(10), 627–639. Advance online publication. 10.1016/0950-5849(92)90068-Z

Hilburn, T. B., Hirmanpour, I., Khajenoori, S., Turner, R., & Qasem, A. (1999). *A Software Engineering Body of Knowledge Version 1.0*. Academic Press.

How To Become A DevOps Engineer In 2022: A Comprehensive Guide. (n.d.). Retrieved December 8, 2022, from https://devopscube.com/become-devops-engineer/

Khan, F., Kumar, R. L., Kadry, S., & Nam, Y. (2021). The future of software engineering: Visions of 2025 and beyond. *Iranian Journal of Electrical and Computer Engineering*, 11(4), 3443. Advance online publication. 10.11591/ijece.v11i4.pp3443-3450

Leite, L., Rocha, C., Kon, F., Milojicic, D., & Meirelles, P. (2019). A survey of DevOps concepts and challenges. In *ACM Computing Surveys* (Vol. 52, Issue 6). 10.1145/3359981

Martínez, Y. V., Domínguez, E. L., Velázquez, Y. H., Isidro, S. D., Medina Nieto, M. A., & de La Calleja, J. (2019). Layered software architecture for the development of third-generation video surveillance systems. *IEEE Access : Practical Innovations, Open Solutions*, 7, 98507–98521. Advance online publication. 10.1109/ACCESS.2019.2930401

Moreb, M. (2022). *Practical Forensic Analysis of Artifacts on iOS and Android Devices*. Practical Forensic Analysis of Artifacts on IOS and Android Devices. 10.1007/978-1-4842-8026-3

Paolone, G., Marinelli, M., Paesani, R., & di Felice, P. (2020). Automatic code generation of mvc web applications. *Computers*, 9(3), 56. Advance online publication. 10.3390/computers9030056

Shankar, P., Morkos, B., Yadav, D., & Summers, J. D. (2020). Towards the formalization of non-functional requirements in conceptual design. *Research in Engineering Design*, 31(4), 449–469. Advance online publication. 10.1007/s00163-020-00345-6

Tiwari, S., & Singh Rathore, S. (2019). Teaching Software Process Models to Software Engineering Students: An Exploratory Study. *Proceedings - Asia-Pacific Software Engineering Conference, APSEC*. 10.1109/APSEC48747.2019.00049

Chapter 3
Requirements Engineering

ABSTRACT

This chapter focuses on the crucial first stage of software development: requirements specification. This stage serves as a formal agreement between the developer and the client, outlining the software system's purpose and functionality. The requirements specification provides a detailed description of the system, addressing the client's needs and expectations. It encompasses the system's capabilities (what it does and doesn't do), user interactions, quality requirements, and any design constraints. Through this process, readers will gain the knowledge and skills to select and justify appropriate tools and techniques for realizing a custom-built website during the subsequent design phase.

BACKGROUND

examination system used by many universities, we introduce about Abc online examination system (AOES) system and compare it with a current manual system that is applied by universities, most education trends push and advise education institutions to apply smart learning in classrooms, in the next sections we will analyse the system requirements functional and nonfunctional requirements of each of requirements will module and determine how to interact with other modules which relation and type or each relation, using uniform modelling language (UML) to describe objectives that AOES system includes. We describe in detail how analysis system requirements include Use Case Diagram, Use Case Priority List and Use

DOI: 10.4018/978-1-6684-4844-1.ch003

Case Specifications then we describe Non-functional requirements for the system in chapter three.

The purpose of this chapter is it propose all features and procedures to develop the system for Abc University that will apply to English teachers. In this chapter we will study and introduce the current system, analyse the requirements, and determine the functional requirements for objects for the system, the UML 2.0 language (*About the Unified Modeling Language Specification Version 2.0*, n.d.) will be used to analyse and describe our system to determine the details of objectives, scope, design model, requirements and then we will study and monitoring, reporting mechanisms that will apply for developing the system. The motivation of our practical example as this chapter as this example is designed for the similar tasks that we want to create, but they exist in a general formula i.e. they should played, and commercial that cannot be customized as Abc University needs, in addition, there are some features in them also the Abc university does not need. Some of these web-based systems will be as following in short:

- Examino - Online Exam System (*Examino - Online Exam Maker*, n.d.), Examino is an online exam creator script used to build Online Exam online. Using Examino as an assessment system to build exams with several features, such as example multiple question types, Paid Exam system, Exam result system.
- USATestprep (*USATestprep: K-12 Standards-Aligned Practice Tests*, n.d.) Integrates district reporting, assessments, individualized learning and classroom support to improve student achievement for elementary. Offering a multitude of test review products for subject area end-of-grade, end-of-course, graduation level and college entrance assessments.
- Steve Penfold (Vishwas Kotwal et al., 201introduceduce about most important eLearning Trends To Watch In 2017 and using Microlearnineducationiothe n the is most important to apply. It provides a structure to deliver learning content in a way that fits with what we know about how people learn - short, focused, regular challenges that stretch and, perhaps, reward us. Many moderuse thisusethis concept in learning apps, such as Duolingo.

Statement of Problems in the Current System

This exam will be introduced to a student who is go to take to make test in specific date and time even the university determines its level after that an answer of student will be taken, and checked then put all the results of all students in mark sheets, finally the student will be informed by an announcement board or by e-mail, in addition it's level will be known to the university.

all the exam process doing by paper based, this needs a lot of time, the whole process of assigning test and evaluating their scores after the test, was done manually until date. It is very time wasting, also it is difficult to keep the answer sheets being generated as well as the maintenance of the record of each examination. The chance of losing any information about such a student represented in its grade is high and searching operation is difficult. Result processing takes more time and the presence of more invigilators operation that can be done particularly if there are a lot of students in the exam, as an instance if the university has 500 as new students how they need abut 500 * 5 papers = 2500 papers and 500 students / 25 students per a class = 20 room to hold class, 25 teachers for exam management process and correct answer.

New System

Abc online examination system (AOES) can be used by any teachers in the department of English language for developing their strategy from paper base to electronic system managed by AOES for putting the exams and reduces time in managing and prepare exam.

AOES can summarize in simple steps for teacher, student, and administrator as users will open the online system, this system will install and configure on a server for university and admin by administrator, after student login to the system he/she can open Exam that a teacher prepared it before, this exam has rules added by teacher to open at specific time and date, when the student submit all questions, the system will save the student results and display to student directly, after a session finished a teacher or an admin can download all student results from the system.

This AOES System is a software solution, which allows any industry or institute to arrange, conduct and manage examinations via an online environment. (Vishwas Kotwal et al., 2015), AOES is a system that Abc university English teacher and all student in university can benefit from it. Current system in university that use paper-based materials and pens to process the manual examination. But AOES system provides the student information, exam questions with answers and automatically calculate the scores. Users can access the system from university website. After student opens its exam the system will open the exam for a limited time to answer the questions and after the time expiry all answer will submitted to system and end student session automatically, and all students answers will be logged on the system, Figure 1 describes the main steps that we have just talked about it.

Figure 1. Describes main activity for AOES for system users (admin, teacher, student)

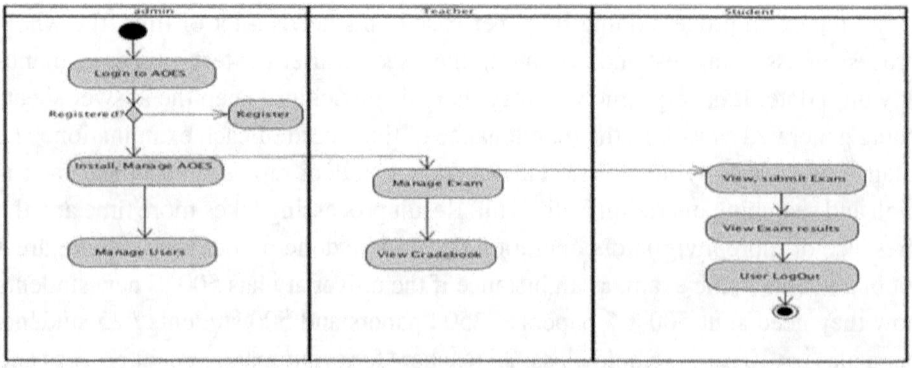

AOES is very helpful to users. This practical example aims to provide quick, immediate and easy way to appear the exam. It can provide special advantages to the students/applicants. The AOES can automatically add the marks allocated in each question to determine the total mark for the questions. A time limit can be set for the questions. AOES allows jumping to specific questions based on the previous answer.

The AOES limits the number of attempts a student can answer a question. Login module helps the user to login to the site. For that he/she must type the username and password correctly. The login provision in this page helps the already registered user to directly access the site and there is a link for registration to a user who is new to this site. Student module is mainly for the students. This helps the students to register for the exam and answer the exam. For registration name, student ID, role id, password should be entered and other are filled. This will provide result after the exam according to correct and wrong answer. Administrator module is mainly for the administrator. This will contain the creation of exam then add questions, preview of already created question paper, and the report of the administrator. For creating the question paper he/she must enter Exam ID, No: of questions, total time (in min), Marks per Question, Option type. For showing the preview of already created question paper he/she must enter the correct Exam id.

the system, mainly related to three categories are: students, teachers and administrators. Students take the exam online, they use user names and passwords the administrators given to them to login, the system will distribute questions to the student automatically, and start the countdown, make the students to complete the test within the stipulated time, and submit the answers to the server. According to the form of the exam, the questions are divided into objective questions and subjective questions, papers are corrected by the system and teachers. Final results are aggregated by the system; students can check as describe in Figure 2.

Figure 2. Operation flowchart of student examination system

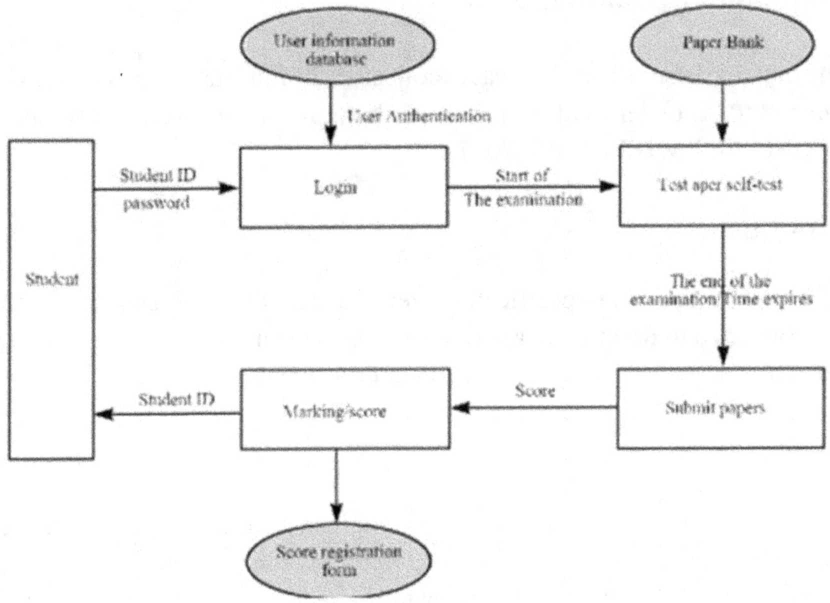

AOES will enhance our exam strategy by include these trends in the when student use the system because each student will have its area to listen to question without noise around him and students will save time that they use to fill answer sheet, for preparing exam questions is challenges, tedious and time consuming for the instructors. AOES helps instructors to store the questions banks in computer databases that allow a teacher to reuse it later in other semester or sessions, AOES will use connection to current DB to validate registered user to login to site, and admin can back up the student marks to registration office to be integrated with university system.

By applying AOES on English teacher at Abc University will reduce the time and efforts and reduce the cost by applying online examination system, students can see their result directly after submit the exam and from interview with staff in English teacher, addition to that AOES will be for English teacher and we can use it for other exam in university but the requirements fit with this documentation.as following:

a- English teacher in Abc University and other teacher can use the system.

b- AOES will available online as web application and user can access it online.

c- Admin can add roles to determine which computer ip can access the exam.

SYSTEM LIFE CYCLE

Requirement Specification

The first step in AOES project was to collect all the information about the existing system and try to understand the problems that it facing and to find a solution for it (ISO/IEC 2017 & IEEE 2017, 2017).

Analyzing

After the requirement specification step we start the analyzing phase and in this step we tried to describe our new step by using some UML's diagrams, which included use case diagrams and class and interface diagrams.

Design

After the analysis phase, we start the design step in this step also we used in details UML's diagrams for each case that determine in previous phase, Design is the abstraction of a solution and view patterns seen in the analysis phase to be a pattern in a design phase, which included state diagrams and sequence and activity and some other diagrams that represent the interaction between components and represent stream functioning of the system.

Implantation

The final stage in our project is the implementation step where we have assimilated all classes to the programming language. There have been difficulties especially with the reservation part because we need to consider all the dates that starting from the reservation date until the return date also was a bit difficult to compare the dates, but we have tried to implement as much as possible. The system has been develop using MySQL database and php by using macromedia Dreamweaver to write code and design html page and user interface and our database (Alter, 2001).

FUNCTIONAL REQUIREMENTS

The following points represent the functional requirement (Jang et al., 2021):

- Student can access exam online using computer or smart phone devices.

- English teacher will get students gradebook automatically after the last student submit his attempts on the time end for exam.
- Teacher will add marks for each question in exam and total exam mark will depends on summation of total question marks.
- Responses by the candidates will be checked automatically.
- time consuming, Staff cost and paper work will reduce.
- Teacher can create different common type of question include multiple options, multiple answers, true false, fill in blank.
- Student can get exam result automatically after submitting and finishing exam.

System Functionalities

AOES is web application provide question to the student randomly should include three type of users (student, teacher and administrator) as describe below, the main function for the system to create the exam, display to students to fill and submit and get result, exam timer should be show once the student start Exam attempts, the system should satisfy the three sub modules, Figure 3 summarize the user functions that they use, admin will have use all the functions that available to students and teacher, teacher can use student function after admin allow or disallow in user roll table, student can access the system and start his exam (Benfell, 2021).

Figure 3. User requirements with functional requirements that will use by users

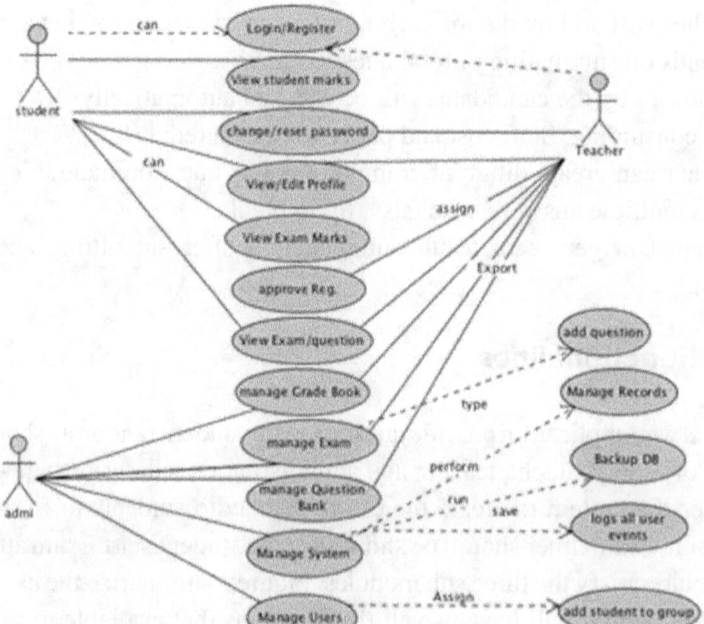

System User Requirements

Student Module

1. Can access registration form.
2. Logging into the system.
3. View exam with questions.
4. Appearing for the teacher/admin.
5. Can change the password.
    7. Resetting of forgotten password.
    8. View/modify profile.

Administrator Module

1. Back up for the database.
2. Full control to manage the system.

3. Create/delete/manage/update/view user account.
4. Editing/Deleting/Creating the records.
5. Add teacher.
6. Can change the password.
7. Log all event for user.
8. Record/manage grade book for all students with attempts they apply.
9. System should support WAN so that exam can given from anywhere.

Teacher Module

1. Logging into the system.
2. Set question marks, time limit.
3. Accepting registrations of student.
4. Adding/editing/deleting different question type (multiple, True/False, Fill text).
5. Adding student to a group/class.
6. Adding/editing/deleting exam.
7. add correct answer for question.
8. View student marks/attempts.
9. Change password.
10. View/modify profile.
11. Declare exam schedule and Results.
12. Manage Question Bank

System Module

1. Auto backup system.
2. Time management enable/hide exam.

Specific Requirements

In this section, we try to analyze the overall activity of the AEOS system by Use Case Diagram, Use Case Priority List, and Use Case Specifications.

Actor Description

Actors: are external entities that interact with the system. Actor initiates system activities to complete some task.

So actors in this project are as follows: -

Administrator: Super user, adds teacher and manages system, do all admin task and access all system feature.

Teacher: Another user name as teacher member, lecturer or examiner who create exam and questions with answers and can access grad book.

Student: User mostly a student who will appear for the examination.

System: the system do automatic task that assign by user and should execute depend on time or events.

Use Cases

A use case describes a sequence of interactions between a user and system, without specifying the user interface. Use cases describe the system functions that describe previous section.

List of Use Cases Are

- Login
- Student registration
- Add teacher
- Add student to group
- Add Exam
- Add question
- Get Exam Marks
- Backup database
- Restore system
- Logout

User Use Case Diagram

Figure 4 describes a User Use Case Diagram for our example as it depicts three user roles: Administrator, Teacher, and Student. Each role interacts with the system through specific functionalities. The Administrator can add departments and log in, the Teacher can manage student groups, register students, and add questions, while the student can appear for tests and benefit from result generation. The diagram illustrates how these users interact with the system to perform their respective tasks.

Figure 4. User use case diagram

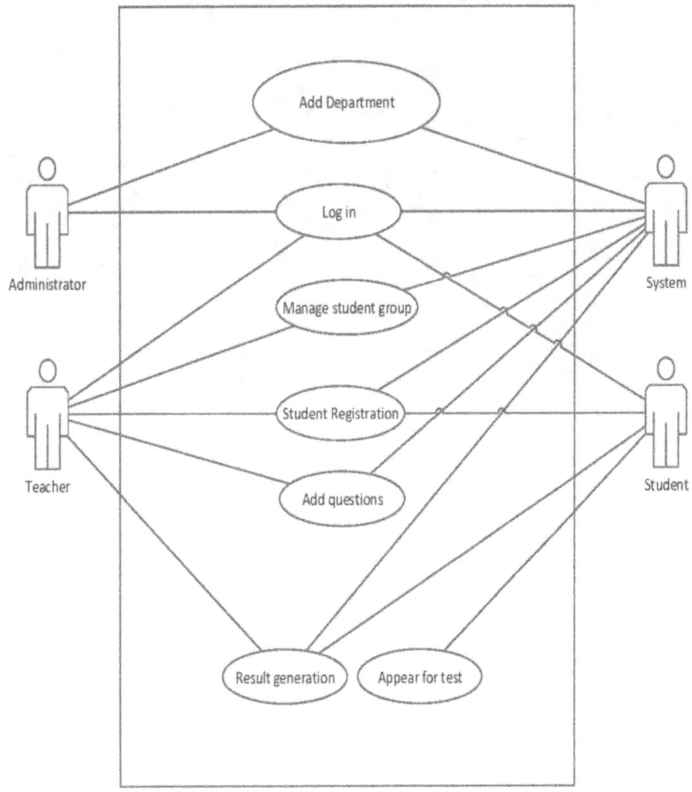

Add Question Use Case

a use case diagram for the process of a teacher adding a question to an educational system shown in Figure 5. The diagram outlines the steps involved, starting with the teacher logging into the system. They then set the course name, enter the question, and select the choice type. The teacher can choose to select a group for the exam and set the correct answer for the question. The diagram also highlights the inclusion of sub-functions such as setting question options, allowing user answers, and incorporating timed questions with negative mark ratios. All these functions are connected to the overall system, showcasing the integrated process of adding a question.

Figure 5. Add question use case

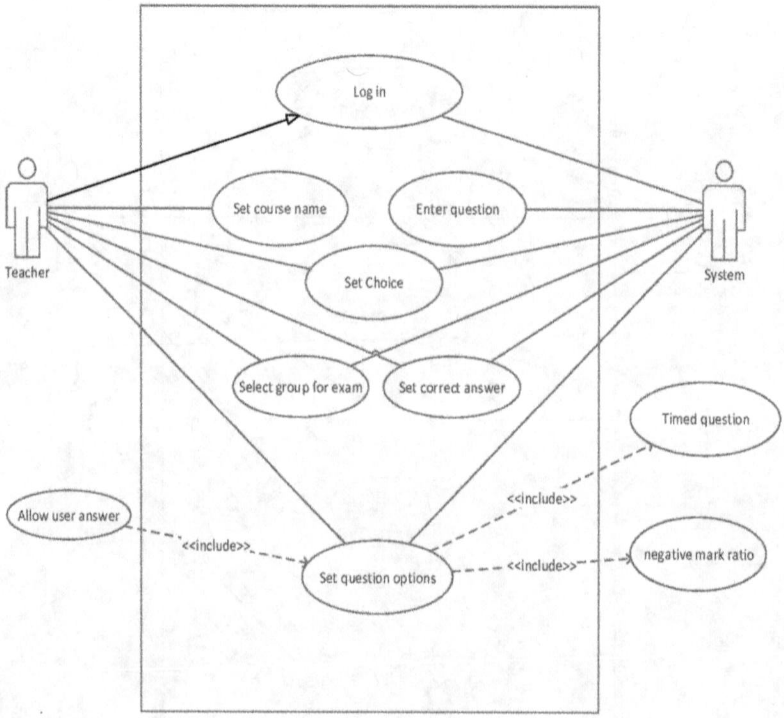

Start Exam Use Case

A user requirement is for the student to be able to select between different question types, as indicated by the "Select question type" action. The student starts by logging into the system, then selects a question and chooses the question type. The answering process includes both "True/False" and "multiple choice" question types. A system requirement is the ability to handle different question formats, ensuring the system can present and process both "True/False" and "multiple choice" answers. After answering the selected question, the student can either logout from the system or proceed to select another question and repeat the process. This use case demonstrates the basic flow of a student taking an online exam, highlighting the key actions and interactions involved. Figure 6 illustrates the "Start Exam Use Case" for a student interacting with an online exam system.

Figure 6. Start exam use case

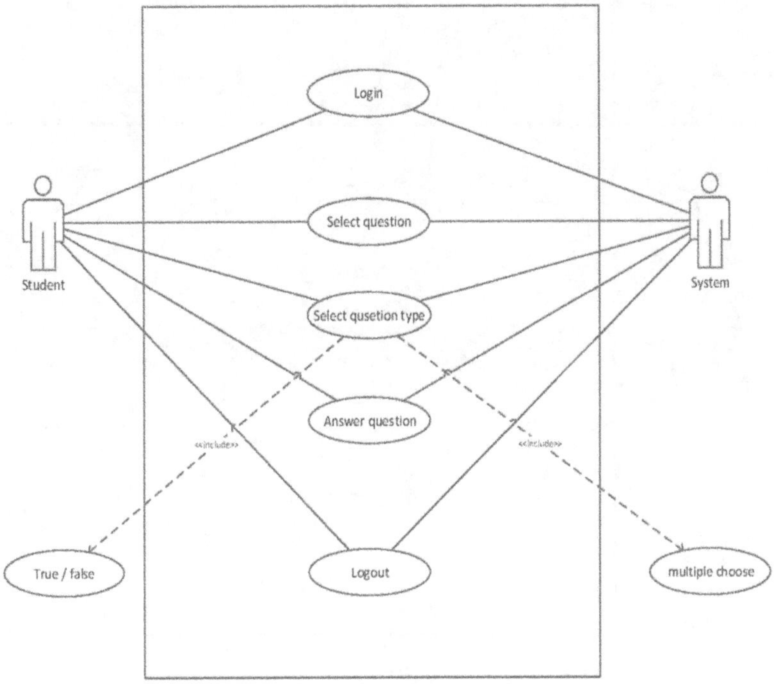

Calc. Exam Result Use Case

This use case focuses on the system's process for calculating exam results. The system checks answers, calculates marks, and sets the final marks for the exam. Included in this process is the counting of both correct and incorrect answers. A system requirement is the ability to calculate marks based on a formula that considers the number of correct answers, the number of wrong answers, and a pre-defined negative mark ratio. The formula (Marks= no. of correct answer) is outlined within the diagram. This use case highlights the automated calculation of exam results, ensuring accurate and efficient marking.

Figure 7. Calc. exam result use case

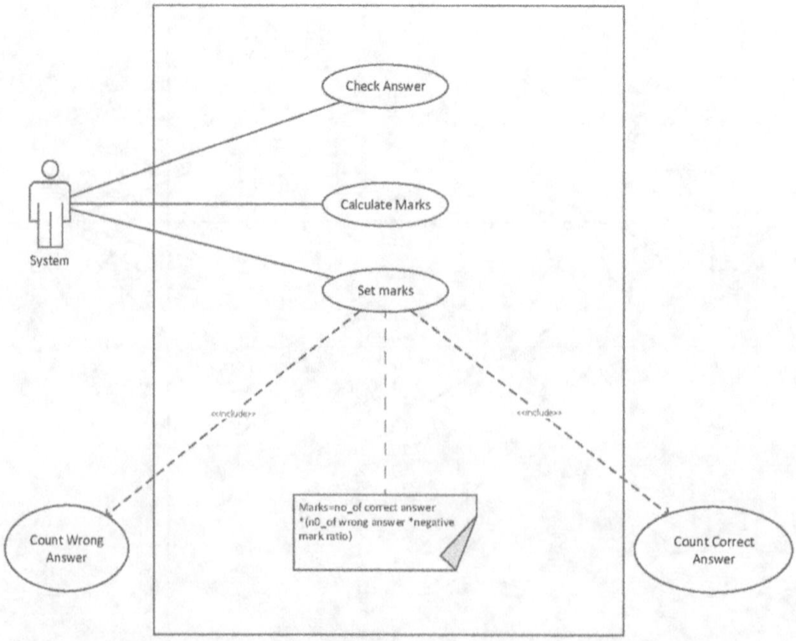

Login Use Case

login process for our example used in this chapter can be describe as shown in Figure 8 which describe the interaction between various user roles including Administrator, User, Teacher, and Student with the system's login functionalities such as Login, Logout, ChangePassword, and UpdateProfile. A key user requirement is the ability for all authorized users to log in using their unique username and password. The detailed steps of the login process are outlined, starting from the user wanting to login to the system, entering their credentials, system verification, and final acceptance. Alternative flows for scenarios like a user having no account or entering incorrect credentials are also addressed. This comprehensive use case clarifies the functionalities and expected user interactions within the AEOS-LOGIN package, ensuring a clear understanding of the login process.

Figure 8. Use-case diagram for the AEOS -LOGIN package

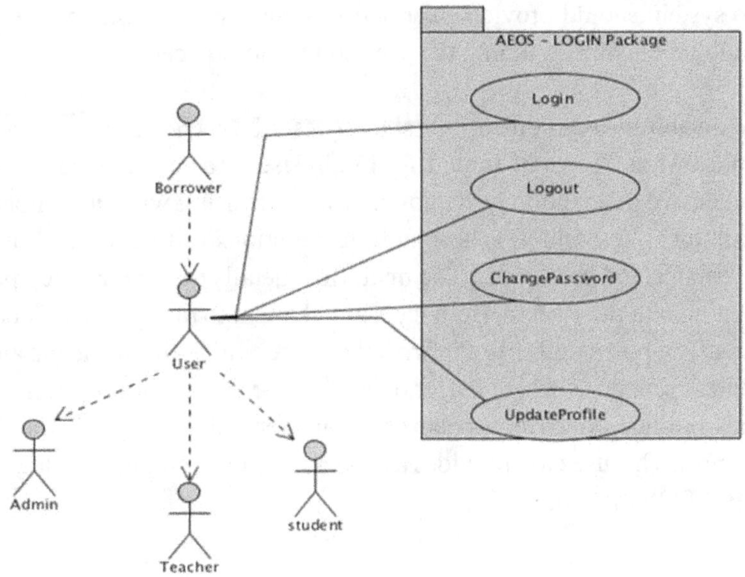

System requirements and user requirements ensure that the Login package provides a secure and user-friendly login experience, facilitating authorized access to the system while safeguarding user credentials and allowing for account management, each requirements can be describe as below:

System Requirements:

- The system must be able to authenticate users based on their unique username and password and validate entered credentials against stored data.
- The system must allow for the creation of new user accounts and support functionalities for users to change their passwords and update their profiles.
- The system needs to securely store user credentials (usernames and passwords) and handle incorrect login attempts gracefully, potentially with lockout mechanisms after multiple failed attempts.

User Requirements:

- All authorized users (Administrator, User, Teacher) must be able to log in to the system using their designated username and password.
- Users should be able to create their accounts with unique usernames and passwords.

- Users must be able to change their passwords.
- Users should be able to update their profiles.
- The system should provide clear and informative messages in case of incorrect login attempts, guiding the user on how to proceed.

Based on above Requirements we should describe each use case include actors, condition, flows as shown in Table 1 for Login Use case describe table. It provides a comprehensive description of the login process, starting with the primary actors (Administrator, User, and Teacher) and preconditions (all authorized users must have a username and password). The main flow details the step-by-step process of a user logging into the system, from their initial intent to the system accepting their credentials. The post-condition confirms the successful login, while the alternative flows address scenarios where a user does not have an account or enters incorrect credentials, outlining steps for account creation, user addition to groups, and retrying login attempts. This use case provides a clear and concise guide to the login process within the AEOS system.

Table 1. Login use case

Use Case: Login Use Case	
ID	**AEOS- LOGIN-01**
Brief Description	Login into the system.
Primary Actors	Administrator, User and Teacher.
Secondary Actors	None
Preconditions	All authorized has the user name and password
Main Flow	1. All authorized user wants to login into the system. 2. User first must have username and password. 3. User enters their user name and password. 4. The system checks the user name and password. 5. The system accepts user name and password. 6. The use case ends
Post condition	The system accepts user name and password.
Alternative Flows	**action A:** if the users have no user name and password. 1. A.2 Users create their own new account. 2. A.3 teacher/ admin add user to group. **action B:** if the entered user name and password is wrong. 1. B.3 Try with another user name and password.

Update profile use case describes in detail how a logged-in user, primarily the Admin but also potentially Teacher or Student, can change their profile information. The main flow outlines the steps involved, from clicking the "Update Profile" button to successfully changing the profile. The post-condition specifies successful profile update. Alternative flows address situations where the user enters incomplete data or encounters errors, including potential issues with the "Change Password" function, and highlights the system response with error messages for incorrect entries. This use case provides a clear guide for users to understand and navigate the process of updating their profiles within the AEOS system as described in Table 2.

Table 2. Update profile use case

Use case: Update Profile	
ID	**AEOS-LOGIN-04**
Brief Description	The User changes his or her Profile Information.
Primary Actors	Admin
Secondary Actors	Admin, Teacher, Student
Preconditions	The Use-Case starts when: The User is currently logged on the System. The User is navigating on the main menu.
Main Flow	The User clicks 'Update Profile'. The System displays the 'Profile Page' and "update" Button at the top of the page. The User click on "update" add or fill profile fields then clicks 'Update Profile. The System successfully changes the Profile. [ErrorProfileData]
Post condition	The Use-Case terminates when: • The User successfully update profile page.
Alternative Flows	[ErrorProfileData] • The User enters incomplete profile data 'Change Password'. • The System displays an error box telling the profile field error entered is incorrect.

logout use case describing the process of a user logging out of the system, it clearly defines the expected user interaction and system response for the logout functionality within the AEOS system. Primarily used by the admin but also applicable to Teacher and Student roles, the use case starts with the user being logged in and navigating the main menu. The main flow describes the two-step process: the user clicking "logout" and the system subsequently logging them out. The post-condition confirms the successful logout. No alternative flows are identified, suggesting a straightforward logout process with no anticipated errors or deviations as described in Table 3.

Table 3. Logout use case

Use case: logout	
ID	**AEOS-LOGIN-03**
Brief Description	The User logs out of the System.
Primary Actors	Admin
Secondary Actors	Admin, Teacher, Student
Preconditions	The Use-Case starts when: • The User is currently logged on the System. • The User is navigating on the main menu.
Main Flow	1. The User clicks 'logout'. 2. The System logs the User out.
Post condition	The Use-Case terminates when: • The User successfully logout.
Alternative Flows	None

The use case for "Change Password" describes the process for a logged-in user to change their current password, with the admin being the primary actor and Teacher and Student as secondary actors as described in Table 4. The main flow details the steps involved, starting with navigating to the "Change Password" menu, entering the old and new passwords, confirming the new password, and the system successfully making the change. The post-condition highlights successful password change. Alternative flows address the scenario of an incorrect old password being entered, where the system prompts the user with an error message indicating the incorrect entry. This use case provides a clear understanding of the expected user actions and system responses for changing passwords within the AEOS system.

Table 4. Change password use case

Use case: change Password	
ID	**AEOS-LOGIN-02**
Brief Description	The User changes his or her current password.
Primary Actors	Admin
Secondary Actors	Admin, Teacher, Student
Preconditions	The Use-Case starts when: The User is currently logged on the System. The User is navigating on the main menu.

continued on following page

Table 4. Continued

Use case: change Password	
ID	**AEOS-LOGIN-02**
Main Flow	The User clicks 'Change Password'. The System displays the 'Change Password' menu. The User enters its old password, new password and confirm new password then clicks 'Change Password'. The System successfully changes the password. [Incorrect old password]
Post condition	The Use-Case terminates when: • The User successfully changes its password.
Alternative Flows	[Incorrect old password] • The User enters his old password and new password and clicks 'Change Password'. • The System displays an error box telling the User that the old password entered is incorrect.

Admin Use Case

Admin use case outlining the system and user requirements with two key actors are involved: the student, representing individuals taking exams, and the Admin, responsible for managing the system, the diagram is structured into two packages: User Package and System Package. The User Package focuses on functionalities related to student interactions. "Manage User" allows admins to create, edit, and delete student accounts. "EditUserDetails" enables the modification of student information. "Add Teacher" enables the addition of new teachers to the system, while "Add Student" allows the addition of new students.

The System Package handles the administrative aspects of the online exam system. "Backup" enables admins to manually create system backups, with "CreateBackup" representing the process itself. "AutoBackup" facilitates automated data backups. "Grade book" provides a platform for managing exam scores and student performance. "Manage" encompasses overall exam management, including exam creation, scheduling, and distribution. "Log all event" records all student and admin activities within the system. "Record" allows admins to log various data points related to exam administration. "Manage Records" enables the management and organization of recorded data. Lastly, "Time management" allows admins to manage time-related aspects of the exam system, such as setting exam durations and deadlines.

Figure 9 underscores the essential requirements of admin use case, as students need to be able to access exams, submit answers, and view their results. System requirements include user account management, exam creation and management, secure exam delivery, automated grading, result analysis, data security, system reliability, scalability, and ease of use. Additional considerations may include integration with existing learning platforms, exam question randomization, plagiarism detection, and robust reporting capabilities for exam analysis.

Figure 9. Admin use case package

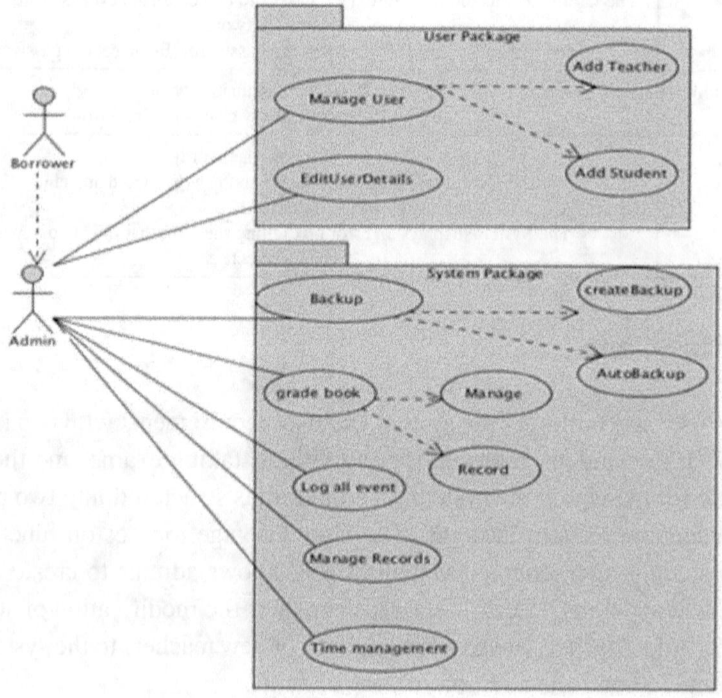

Add Teacher use case as describe in Table 5 highlighting the necessary steps, involved actors, preconditions, main flow, post condition, and alternative flows, as shown it outlines the steps involved when an administrator adds a new teacher to the system, the primary goal of this use case is to add a new teacher to the institution's database. The administrator is the primary actor responsible for executing this process, while no secondary actors are involved. Before initiating this process, the administrator must have access to the teacher's code and name. This requirement serves as a precondition for the use case. The main flow of the use case begins with the administrator's intention to add the teacher to the database. Subsequently, the administrator retrieves the teacher's code and name, this information is then used to add the teacher to the system, utilizing both their name and code. Once the teacher is successfully added, the use case is considered complete.

The post condition indicates that the teacher is successfully added to the system. However, an alternative course of action exists: if the teacher doesn't have a code, the administrator will assign one, this alternative flow ensures that all teachers are properly registered within the system, even if they initially lack a designated code.

Table 5. Add teacher use case

Use Case: Add Teacher	
ID	**AEOS-Admin-01**
Brief Description	Adding new teacher to the institution.
Primary Actors	Administrator
Secondary Actors	none
Preconditions	The administrator first knows the teacher code and teacher name.
Main Flow	1. Administrator wants to add teacher in to database. 2. The Administrator looks the teacher code and teacher name. 3. The Administrator adds the teacher by name and code of the teacher. 4. Teacher is added. 5. The use case ends
Post condition	Teacher is added.
Alternative Flows	**Alternative course of action A:** if the teacher has no code The administrator gives it a code.

Result Generation use case highlighting the steps, actors, preconditions, main flow, post condition, and alternative flows, providing a comprehensive understanding of the result generation process within the online exam system, includes detailing the process of generating exam results. it involves both the teacher and the system as primary actors, while no secondary actors are involved. The primary purpose of this use case is for teachers to generate results by calculating submitted answers from students.

Before the process begins, the system needs to have received all the submitted answers. This acts as a precondition for the use case. The main flow of the process involves several steps. Initially, the system retrieves the submitted answers. It then checks each answer and counts the number of correct and incorrect answers, based on these counts, the system calculates the overall mark. The teacher then posts the calculated results, officially generating the results. Finally, the results and student numbers are saved to the system's gradebooks, marking the conclusion of the use case.

The post condition indicates that the results are successfully generated. However, an alternative flow is defined for situations where a posted result might be incorrect. In such cases, the system will re-check the answer, ensuring the accuracy of the generated result.

Table 6. Result generation use case

Use Case: Result Generation	
ID	**AEOS-Admin-02**
Brief Description	Teacher generates the result by calculating the submitted answer by student.
Primary Actors	Teacher and System.
Secondary Actors	None
Preconditions	System gets submitted answers.
Main Flow	1. System gets submitted answers. 2. System checks answer. 3. System counts correct answer 4. System counts wrong answer. 5. The system calculates the mark. 6. Teacher posts the results. 7. Result is generated. 8. Result and student number saved to gradebooks. 9. The use case ends.
Post condition	Result is generated.
Alternative Flows	A: if the posted result is wrong A.2 The system checks the answer.

The Create Backup use case as describe in Table 8 detailing the steps, actors, preconditions, main flow, post condition, and alternative flows, providing a comprehensive understanding of the database backup process within the online exam system, it details the process of manually backing up the system's database by the administrator. It's identified as AEOS-ADMIN-03, serving as a unique identifier. The primary actor in this use case is the administrator, with no secondary actors involved.

Before the backup process can begin, the administrator must be logged into the system and navigate through the admin menu. These are the preconditions for initiating the use case. The main flow begins with the administrator clicking on the "Create Backup" option. The system then displays a confirmation screen asking the administrator if they want to create a backup. The administrator confirms this action, leading the system to create a folder named with the current date and time. Then, the system copies the entire database into that folder. However, if there isn't enough space, the backup process might fail.

The use case concludes successfully when the database has been successfully backed up. This is the post condition. There are two potential alternative flows. If there's insufficient space, the system will prompt the administrator with an error message indicating that the database can't be backed up due to a lack of space on the selected drive. Alternatively, if the backup process fails, the system will prompt the administrator with an error message that the backup has failed.

Table 7. Create backup use case

Use Case: Create Backup	
ID	**AEOS-ADMIN-03**
Brief Description	The Admin backs up the database manually.
Primary Actors	Admin
Secondary Actors	
Preconditions	The Use-Case starts when: • Admin is logged on to the system. • The admin is navigating through the admin menu.
Main Flow	1. The Admin clicks on 'Create Backup'. 2. The System displays a confirmation screen asking the Admin whether he is sure he wants to backup 3. The Admin confirms the creation of a Restore point. 4. The System creates a folder named with the current date and time and copies the entire database into that folder. *[Not enough space] [Backup failure]*
Post condition	The Use-Case terminates when: • The database has successfully been backed up.
Alternative Flows	*[Not enough space]* • The System prompts the Admin through an error message that it cannot backup the database, as there is not enough space on the selected drive. *[Backup failure]* • The System prompts the Admin through an error message that the backup has failed.

Auto Backup use case as described in Table 8 outlines the automated process of backing up the system's database every day, it's identified as AEOS-ADMIN-04. Both the administrator and the system itself are primary actors in this use case, with no secondary actors involved. The automated backup process is initiated at 18:00 every day. The system creates a folder named with the current date and time and copies the entire database into that folder. However, if there's insufficient space, the backup process might fail. After a successful backup, the system logs the date and time of the event.

The use case concludes successfully when the database has been automatically backed up. However, there are two potential alternative flows. If there's insufficient space, the system will alert the administrator that the backup couldn't be completed due to a lack of space on the selected drive. Alternatively, if the backup process fails, the system will alert the administrator that the backup has failed.

Table 8. Auto backup use case

Use Case: Auto Backup	
ID	**AEOS-ADMIN-04**
Brief Description	The database is backed up automatically every day.
Primary Actors	administrator and System
Secondary Actors	
Preconditions	The Use-Case starts when: It is 18:00
Main Flow	1. The System creates a folder named with the current date and time and copies the entire database into that folder. *[Not enough space] [Backup failure]* 2. *The system save AutoBackup_date_time to log event.*
Post condition	The Use-Case terminates when: • The database has successfully been backed up automatically.
Alternative Flows	*[Not enough space]* • The System emails the admin that it cannot backup the database, as there is not enough space on the selected drive. *[Backup failure]* • The System emails the Admin to alert him that the backup has failed.

Edit User Details use case as described in Table 9 outlines the process of modifying user details by the administrator. It's identified as AEOS-ADMIN-05, serving as a unique identifier. The primary actor in this use case is the administrator, with no secondary actors involved, before the editing process can begin, the administrator must be logged into the system and navigate through the admin menu. These are the preconditions for initiating the use case. The main flow begins with the administrator clicking on the "Edit Current User" option. The system then displays the "Edit Current User" menu. The administrator enters the user ID of the user account and clicks "Find." (It's suggested that finding the user can be done using the first name.) The system then locates and displays the user's details in text boxes. The administrator edits the details they wish to change and clicks the "Done" button. The system successfully changes the details and returns to the admin menu.

The use case concludes successfully when the user's details have been changed. This is the post condition. However, there's an alternative flow for situations involving invalid input. If the administrator tries to submit the page with invalid input in one of the boxes, the system will display an error message prompting the user to re-enter the user details. Alternatively, if the administrator enters a user that doesn't exist in the system, a popup message will prompt them to try again. If the administrator enters the first name of the user account and the user can't be found, the system will display all user accounts with that first name. The administrator can then select the correct user, and the system will bring up the selected user's details ready for editing.

Table 9. Edit user details use case

Use Case: Edit User Details	
ID	**AEOS-ADMIN-05**
Brief Description	The Admin changes the customer details.
Primary Actors	Admin
Secondary Actors	None
Preconditions	The Use-Case starts when: • An Admin is currently logged on the system. • The Admin is navigating through the admin menu.
Main Flow	1. The Admin clicks on 'Edit Current User'. 2. The System displays the 'Edit Current User' menu. 3. The Admin enters the User ID of the user account and clicks 'Find'. *[Find using first name]* 4. The System finds and displays the customer details in text boxes. *[Unable to find user]* 5. The Admin edits the details he wishes to change and clicks the 'Done' button. *[Invalid Input]* 6. The System successfully changes the details and returns to admin menu.
Post condition	• The Use-Case terminates when: The User details have been changed.
Alternative Flows	[Invalid Input] • The Admin tries to submit the page with invalid input in one of the boxes. • The System displays an error message prompting the user to re-enter the user details. [Unable to find user] • The Admin enters user details that could not be found. • The System displays a popup prompting the Admin that the user could not be found. [Search using first name] • The Admin enters the first name of the user account. [Unable to find user] • The System displays all user accounts with the first name. • The Admin selects the correct User. • The System brings up the selected User's details ready to edit.

Exam Use Case

Exam use case outlining the functionalities for both teachers and students within an online exam system with two primary actors represented: the "Teacher" and the "Student. Figure 10 highlights the actions that teachers can perform within the system. "Prepare Question Bank" allows teachers to create and manage a question bank, which includes setting question types and creating individual questions. This use case is extended by "Set Question Type" and "Create Question", "Create Exam" allows teachers to create exams, including adding questions and setting exam details. This use case is extended by "Add Questions" and "Set Exam Details". The student's interactions with the system are also represented. Students can "Login Account" to access the system. "Take Exam" is the primary functionality for students, allowing them to take exams. The "Review Exam Results" use case allows students to view their results after completing exams. The "Register System" use case is included within the "Review Exam Results" use case, suggesting that students need to be registered within the system before they can access their results.

Figure 10. Use case diagram for teacher and student on exam

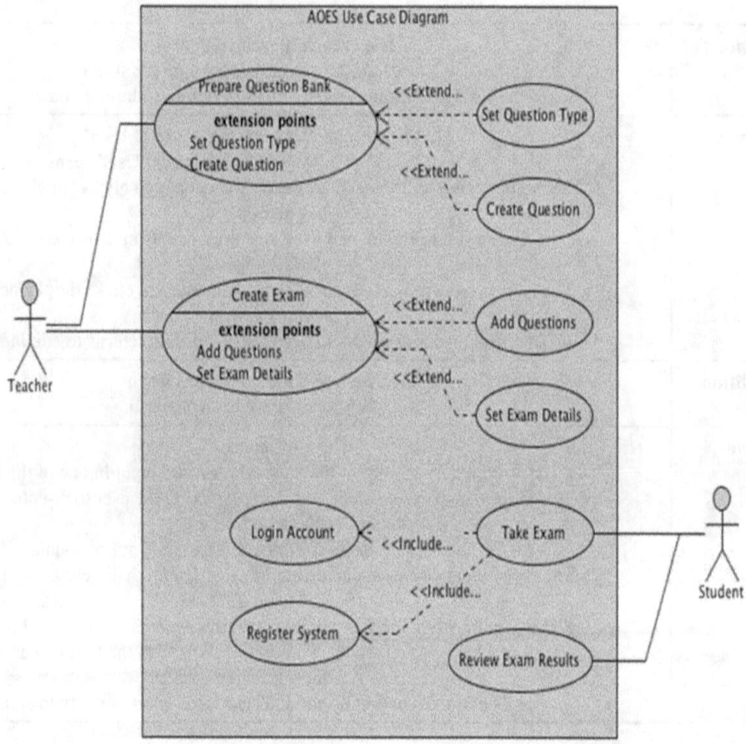

Table 10illustrates the details of when student taking an exam *using* student *actor*, with no secondary actors involved. Before the student can begin taking the exam, they must first log on to the course and be registered to take the exam. They must also be added to the exam group. The exam's start time is equivalent to the time the exam opens. The main flow of the use case begins with the student clicking on the next question to answer, then click on the next question to save their answer and display the next question. The student ends the exam after they have answered all the questions, or the allotted time runs out. The student's results are then saved to the gradebook. The use case is considered complete when the exam has been submitted. However, there's an alternative flow for situations where the student doesn't finish all the questions within the allotted time. In this scenario, the student should submit the answers they have completed.

Table 10. Take exam use case

Use Case: Take Exam	
ID	**AEOS-Exam-01**
Brief Description	The student at specific time can open the exam end start answer question for time limited times.
Primary Actors	Student.
Secondary Actors	
Preconditions	The student first log on to course.
Main Flow	student require to take exam. The student add to the exam group. The time of exam equivalent to open exam time. The student click/select question answer. The student click next question to save answer & display next question. Student end exam after all exam answer or time end. Student result saved to gradebooks. The use case ends.
Post condition	The exam submitted.
Alternative Flows	**A:** if the student did not finish all the given question in given time A.1 The student should have to submit what they answered.

Create Exam use case outlines the process of a teacher creating an exam which identified as AEOS-Exam-02, serving as a unique identifier. The primary actor is the teacher, with the administrator acting as a secondary actor, before the teacher can begin creating the exam, they must first log on to the course and have the intention to add an exam to the course. The teacher will then look up the course name and code, select "Create Exam" from the menu, and add the exam title, date, time, description, and group. The exam is then added, concluding the main flow of the use case.

The post condition as shown in Table 11 indicates that the exam has been successfully added. However, there is an alternative flow for situations where the exam cannot be added. In this case, an error message will be displayed indicating that the exam date and time are not correct.

Table 11. Crate exam use case

Use Case: Crate Exam	
ID	**AEOS-Exam-02**
Brief Description	crate exam according to the courses.

continued on following page

Table 11. Continued

Use Case: Crate Exam	
ID	**AEOS-Exam-02**
Primary Actors	Teacher
Secondary Actors	admin
Preconditions	The teacher first log on to course.
Main Flow	Teacher wants to add exam into course. The Teacher looks the course name and course code. The teacher select from menu Create Exam. The teacher add Exam title, date, time, description, and group. The exam is added. The use case ends
Post condition	The exam is added.
Alternative Flows	**Alternative course of action A:** if the exam can't add. A.1 The exam date and time not correct error massage display.

Add Question use case outlines the process of a teacher adding a question to the system which's identified as AEOS-Exam-03, the primary actor in this use case is the teacher, with no secondary actors involved.

Before the teacher can begin adding a question, they must first log on to the course and open the exam. The main flow of the use case begins with the teacher wanting to add questions to the database. They will then look up the course name and code and review the available options. The teacher will select the question type, add a question name and description, provide the answer for the question, and then add the question by name of the course and course code. The question is added to the question bank, marking the conclusion of the main flow. The post condition indicates that the question has been successfully added. However, there is an alternative flow for situations where the course does not have a code, the teacher will edit the course and add a code to the course as described in Table 12.

Table 12. Add question use case

Use Case: Add Question	
ID	**AEOS-Exam-03**
Brief Description	Adding questions according to the courses.
Primary Actors	Teacher
Secondary Actors	None
Preconditions	The teacher first log on to course and open Exam.

continued on following page

Table 12. Continued

Use Case: Add Question	
ID	**AEOS-Exam-03**
Main Flow	Teacher wants to add questions into database. The Teacher looks the course name and course code. The teacher looks for the options. The teacher select question type. The teacher adds question Name and description. Give the answer for the questions The Teacher adds the question by name of the course and course code. The question is added. The question assign to question bank. The use case ends.
Post condition	The question is added.
Alternative Flows	**Alternative course of action A:** if the course has no code. A.1 The teacher edit course and add a code to course.

View Exam Results Use Case Diagram

Details for view exam result details illustrate the functionalities available to a student within an online exam system it is focusing on the process of taking an exam and accessing their results, with a particular focus on the “View Exam Results” use case. The diagram features a single actor, the “student,” who interacts with various functionalities within the system.

The student can choose to either log in to an existing account or register for a new one (represented by a dashed line) once logged in, the student can modify their profile information, view the exam and its questions, and answer them. After completing the exam, they submit their answers. The “View Exam Results” use case, which is extended by “Submit Exam,” becomes accessible after the exam has been submitted, this allows the student to review their performance.

Figure 11 emphasizes the interconnectedness of these use cases. While the student can optionally log in, register, and modify their profile, the core functionality revolves around taking the exam, answering questions, submitting it, and ultimately accessing their results.

Figure 11. Use case diagram for view exam results

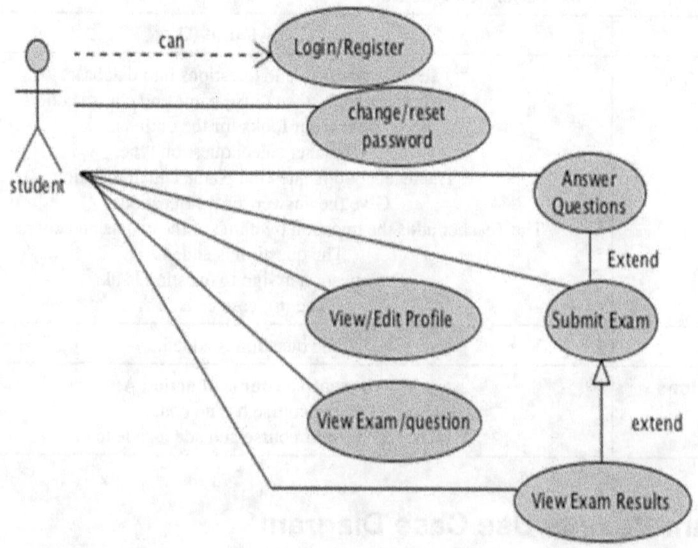

View Exam Results use case details table as describe in table 13 outlines the process of a student reviewing their exam results after submitting the exam. The use case is identified as AEOS-Exam-04, serving as a unique identifier, with the primary actor in this use case is the student, and the teacher acting as a secondary actor.

Before the student can view their results, they must have first taken and submitted the exam and logged on to the course. The main flow of the use case begins with the student opening the exam. The exam results are then displayed to the student, and they can review the answered questions.

The post condition indicates that the results have been viewed. However, there is an alternative flow for situations where the exam is not yet open. In this case, the student will wait until the exam is enabled.

The table effectively outlines the "View Exam Results" use case, detailing the steps, actors, preconditions, main flow, post condition, and alternative flows, providing a comprehensive understanding of the exam result viewing process within the online exam system.

Table 13. View exam results use case

Use Case: View Exam Results	
ID	**AEOS-Exam-04**
Brief Description	The student review exam result after submitted exam.
Primary Actors	Student.
Secondary Actors	Teacher
Preconditions	The student first applies, submitted exam and logon to course.
Main Flow	student open exam. The exam result display to student. The student can review the answered question.
Post condition	The result viewed.
Alternative Flows	**Alternative course of action A:** if the exam not open. A.1 The student waits until exam enable.

Student Use Case

Student use case identified as AEOS-student-01 as shown in Figure 12, the primary actor in this use case is the "student," while no secondary actors are involved. The main purpose of this use case is for students to register with the system. To register, students need to have a unique student ID number. The primary flow of the use case consists of several steps. First, the student accesses the system's website. Then, they fill in all the required information and complete the registration process. Finally, they are registered with the system, and the use case concludes. The post condition indicates that the student is successfully registered, there is an alternative flow for situations where the student does not fill in all the required information. In such cases, the system will display an error message prompting the student to complete the missing information.

Figure 12. Use case diagram for student

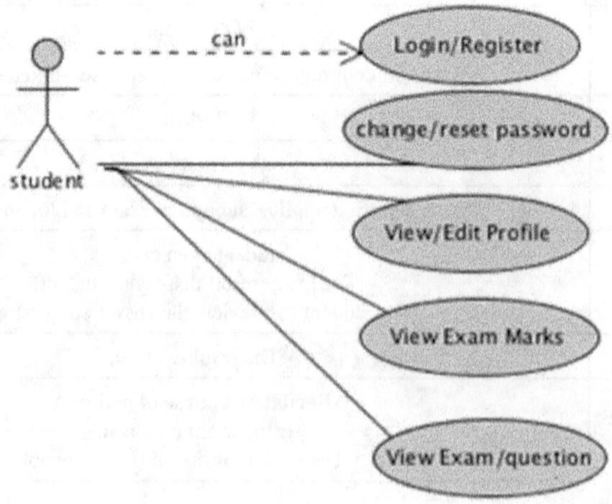

Table 14 provides a clear and concise explanation of the "Student Registration" use case. It outlines the steps, actors, preconditions, main flow, post condition, and potential alternative flows, offering a comprehensive understanding of the student registration process within the online exam system.

Table 14. Student registration use case

Use Case: Student Registration	
ID	**AEOS-student-01**
Brief Description	Student register to system.
Primary Actors	student.
Secondary Actors	None
Preconditions	Students have student id number.
Main Flow	1. Student fill open website URL. 2. Student fill all required student information. 3. Students are registered. 4. The use case ends.
Post condition	Students are registered.
Alternative Flows	**A:** if the student did not fill all their own required information. A.1 the system show error massage.

Teacher Use Case

The Teacher uses case within an online exam system, outlining the process a teacher follows to add students to specific groups with teacher primary actor, and no secondary actors involved.

The precondition for this use case is that students must have sent all their required information, flow of the use case begins with the teacher reviewing the students' submitted information and looking up the student's group. The teacher then fills in the required student information and adds the students to the group as shown in Figure 13.

The post condition indicates that the students have been registered. However, there is an alternative flow for situations where a student has not sent all their required information. In this case, the teacher will announce to the student that they need to send their full information.

Figure 13. Use case diagram for teacher

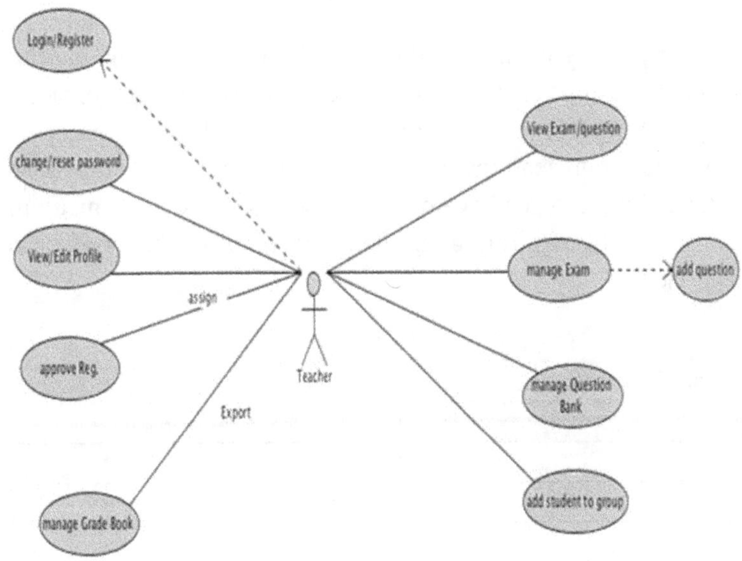

Table 15 outlines the "Add Student to Group" use case, detailing the steps, actors, preconditions, main flow, post condition, and alternative flows, providing a comprehensive understanding of the process for adding students to groups within the online exam system.

Table 15. Add student to group use case

Use Case: add Student to group	
ID	**AEOS-Teacher-01**
Brief Description	Teacher add the students to specific group.
Primary Actors	Teacher.
Secondary Actors	None
Preconditions	Students send all their own required information.
Main Flow	1. Students send all their own required information. 2. Teacher looks for the student information. 3. The teacher groups the students according to their group name. 4. Teacher fill all required student information. 5. Students are registered. 6. The use case ends.
Post condition	Students are registered.
Alternative Flows	**A:** if the student did not send all their own required information. A.1 the teacher announced them to send their own full information.

Use Case Priority List

The use case priority list provided is a helpful tool for organizing the development as it prioritizes use cases based on their importance and impact on system functionality, priority list effectively manage the development process, ensuring that the most important features of AEOS are delivered first, maximizing value and user satisfaction, it logically categorized use cases into high, medium, and low priority as list in Table 16, reflecting their criticality for the system's core value and user experience.

Table 16. Priority rank

Priority rank	Description
1	High priority
2	Medium priority
3	Low priority

High-priority use cases, like login, logout, retrieving exam results, creating exams, managing student groups, and recording results, are essential for the system's basic functionality. Medium-priority use cases, like taking an exam, adding questions, and managing grade books, are important but might be less critical in the initial stages of development. Low-priority use cases, such as changing passwords, updating profiles, and other non-core features, are considered nice-to-haves, potentially added later in the development process.

To refine the priority list, consider specific user needs, the development timeline, the impact of delaying features, and any dependencies between features. For example, prioritize "GetExamResults" above "TakeExam" for students because seeing results is usually a more immediate need. For teachers, "ManageGradeBook" might be more critical than "ViewQuestion" depending on the importance of grade management. For administrators, "RecordResults" could be tied to other features like "ManageGradeBook" and "CreateBackup" to ensure data integrity.

To further refine the prioritization, it's essential to involve stakeholders (teachers, students, administrators) in the process to ensure alignment with their real-world needs. Conducting user research to understand user behavior and preferences will also provide valuable insights. Finally, performing a risk analysis to identify potential risks associated with delaying certain features will help in prioritizing features accordingly as describe in Table 17.

Table 17. Priority use case

Use-Cases – AEOS-LOGIN	Priority
login	1
logout	1
changePassword	3
UpdateProfile	3

Use-Cases – AEOS-Exam	Priority
TakeExam	2
GetExamResults	1
CrateExam	1
AddQuestion	2

Use-Cases – AEOS-admin	Priority
createUserAccount	2
AddStudent	1
CreateBackup	1
AutoBackup	2
RecordResulte	1
ManageGradeBook	2
LogEvents	1

continued on following page

Table 17. Continued

Use-Cases – AEOS-admin	Priority
ManageRecords	1
SetTimeExceed	1

Use-Cases – AEOS-Student	Priority
ResetPassword	2
TakeExam	2
GetExamResults	1
login	1
logout	3
changePassword	3
UpdateProfile	3

Use-Cases – AEOS-Teacher	Priority
addStudentToGroup	2
ViewQuestion	2
GetExamResults	1
login	1
logout	3
changePassword	3
UpdateProfile	2
RecordResulte	1
ManageGradeBook	2
ManageExam	2

NON-FUNCTIONAL REQUIREMENT

non-functional requirements are essential for ensuring the successful development of a robust and reliable for the system analysis, below main points that we should determine for each website to derive non- functional requirement as explain in details in previous chapter (Shankar et al., 2020).

1. Performance

The system must be designed to handle a significant number of users, particularly during peak exam periods. Specifically, it should be able to support concurrent access from 1000 students. Response times are crucial, and the system should respond to user requests (like exam submission, result retrieval, and login) within 3 seconds 95% of the time. The system should also automatically calculate and display student scores based on the correct answers submitted within 5 seconds.

2. Usability

The website interface should be clear, intuitive, and easy to navigate. Consistent navigation, color schemes, and clear instructions will enhance user experience. Furthermore, the system should adhere to WCAG 2.1 Level AA accessibility guidelines to ensure inclusivity. Providing online help documentation and a FAQ section will enhance usability and user support.

3. Portability

The system should be compatible with various operating systems and web browsers. It should run smoothly on Windows 10, macOS 10.15, and Linux Ubuntu 20.04. It should also support major web browsers like Chrome, Firefox, Safari, and Edge. Deployment is crucial; the system should be deployed as a web application on a cloud platform like AWS or Azure for high availability and scalability.

4. Availability

The system should have an uptime of 99.9% to ensure continuous access. Students can only access exams during their assigned time slots but can access other system information at any time. Teachers, on the other hand, should have the ability to register exams and manage exam details at any time.

5. Resolution

The system must accurately calculate and display student scores, providing the correct answers for each question. Error messages should be clear and provide helpful information to users for troubleshooting.

6. Reliability

The system should be designed to minimize errors and downtime. The mean time between failures (MTBF) should be at least 720 hours. The system should maintain daily backups of all user data, ensuring data recovery within 24 hours in case of failure.

7. Integrity

The system should implement robust security measures. Two-factor authentication should be mandatory for all users (students, teachers, and administrators). Role-based access control should be implemented, limiting access to specific functionalities based on user roles. Sensitive data, including student information and exam content, should be encrypted during transmission and storage.

8. Reusability

The system should be modular in design, allowing for easy customization and integration with other educational platforms. This modularity will enhance its re-usability across different universities and institutions.

By focusing on these non-functional requirements, you can create an online exam system that is efficient, user-friendly, secure, and readily adaptable for various educational settings, next non-functional requirement for our case study used in this chapter, which can be documented as:

- Performance:
    - After completing the exam, the entire score of the student will be calculated.
    - The software shall support use of multiple users at a time.
- Usability:
    - The website should be user friendly and should require least effort to operate.
- Portability:
    - The website is develop using HTML, CSS, JSP, PHP, etc. which are platform independent and can be transported to other servers with min-imum effort.
- Availability:
    - Students can take exam only during the previously allotted time slots, however can open site anytime to access other information.
    - teacher can register for the exam anytime.
    - This system must run on multiple operating systems and support win-dows operating system
- Resolution:
    - The system is characterized by accuracy of outputs provided by it is de-scribed accurately in terms of giving the outputs are correct and sound.
- Reliability:

- The system is described as reliable in terms of errors and lack of errors that may affect its operation correctly or disable it for continuous periods of time.
- Integrity:
    - No person who is not authorized to enter this system where the system prevents anyone to enter the system only authorized persons through some of the things that he asks them before entering it.
- Reusability:
    - The system can be reused in all universities and institutes, and not limited to a university.

SYSTEM MODELS

A system model in software engineering is a representation of the software system's structure, behavior, and data. It provides a blueprint for understanding and building the system. In the context of the online exam system (AEOS) example, the system model would encompass elements like user roles (student, teacher, administrator), the flow of data (exam creation, submission, grading, result viewing), and the interactions between different components of the system. For instance, the system model would detail how a teacher creates an exam, how the system stores and manages the exam questions, how students access and take the exam, and how the system calculates and displays results.

This system model would be crucial for various aspects of the development process. It would guide the design of the software architecture, ensuring the system's scalability and maintainability. It would also be used to identify potential bottlenecks and performance issues, allowing for proactive optimizations. Furthermore, the system model would serve as a communication tool for developers, stakeholders, and testers, facilitating a shared understanding of the system's design and functionality.

Object and Class Model

The class diagram for the shown in Figure 14 for AOES package provides a visual representation of the system's structure and relationships between different classes. The diagram reveals a clear separation of concerns, with classes grouped by their roles: student, teacher, admin, exam, question bank, result, and group.

The "Student" class as describe in Figure 14 represents a user who takes exams, with attributes like name, ID, group, and password. The "Teacher" class manages exams, adds students to groups, and views questions. The "Admin" class has administrative control, creating user accounts, managing backups, and recording results. The "Exam" class encapsulates exam-related information like ID, name, duration, and questions. The "QuestionBank" class manages the pool of questions, including their types, choices, and durations. The "Group" class represents a collection of students. The "StudentAttempt" class records student attempts, including response time and selected answers. Finally, the "Result" class stores exam results for each student, including their marks.

Main idea to include class diagram to illustrates the key classes and their relationships, which providing a foundation for understanding the system's internal structure and how data is managed, and it demonstrates the system's functionality through relationships like "Assign" between Student and Group, "Add" between Exam and QuestionBank, and "Submit" between StudentAttempt and Exam. This visual representation is crucial for software developers to understand the system's design and implement its various functionalities accurately.

Figure 14. Class diagram for AOES package

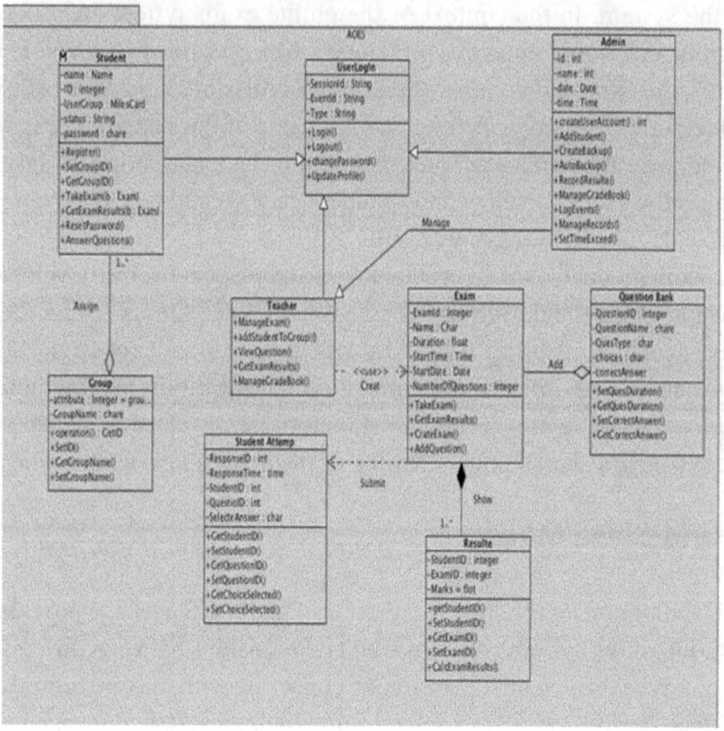

Sequence Diagram

The sequence diagram is a type of interaction diagram used in software engineering to model the interactions between objects in a system over time. Sequence diagrams are particularly useful for visualizing the flow of messages and actions between objects in response to a specific event or scenario. Sequence diagrams are built on a vertical timeline, each object in the system is represented by a vertical lifeline, and messages exchanged between objects are depicted as horizontal arrows across these lifelines. The order of the arrows indicates the chronological sequence of events, making it easy to understand the flow of interactions. Sequence diagrams are valuable tools for communication between developers, stakeholders, and testers. They provide a clear and concise visual representation of how a system will behave, helping to identify potential issues, streamline development, and ensure that all parties have a shared understanding of the system's functionality. This sequence diagram provides a step-by-step visual representation of the interactions between the student and the various system components, depicting the flow of information and actions involved in taking an exam within the system.

In our case, the sequence diagram focuses on the scenario of a student taking an exam in an online exam system, Figure 15 depicts a system-level sequence diagram, illustrating the interactions between a student and various system components when taking an exam, it outlines the steps involved in the exam process. The diagram starts with the "Student" actor, who initiates the process by entering their username and password. This triggers a verification process by the "Login" component. If authentication is successful, the student is authorized. The student then proceeds to select an exam. The "Select Exam" component verifies the student's group and presents the exam. The student then answers the questions within the allotted time, after which they submit the exam. The "Start Exam" component manages the exam timer, while "End Exam" handles the submission process. The "Get Marks" component calculates the student's score, and finally, the "Grade Book" component records the results. Figure 15 highlights the key processes involved in the exam process, including authentication, exam selection, question answering, time management, submission, and result recording. This visual representation aids in understanding the system's behavior and identifying potential dependencies and interactions between different system components.

Figure 15. System-level sequence diagram for student when start the exam

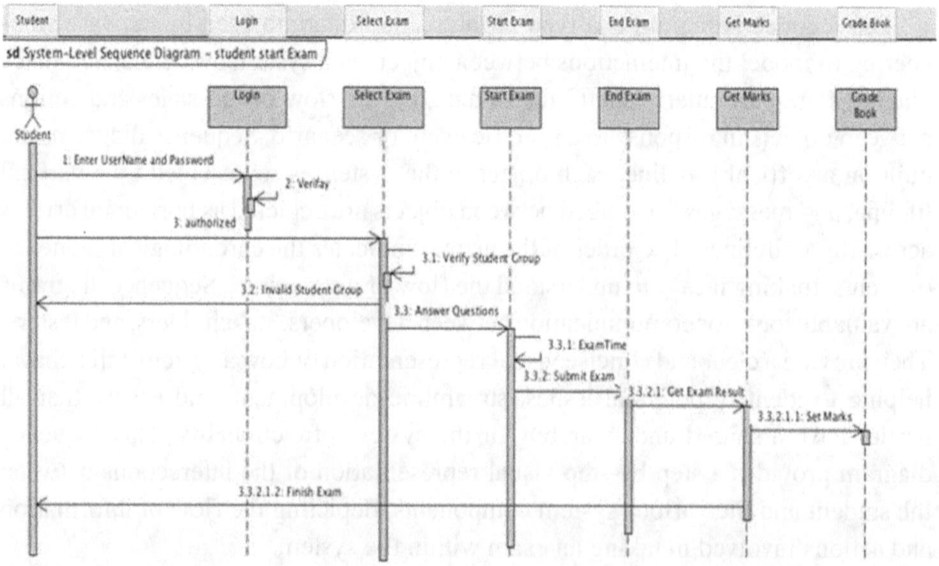

System-Level Sequence Diagram for Add Question and Exam by Teacher

The sequence diagram for this part describes in details the interactions involved when a teacher adds a question to an exam within an online exam system. Figure 16 outlines the steps involved in the process, it starts with the "Teacher" actor, who initiates the process by entering their username and password. This triggers a verification process by the "Login" component. If authentication is successful, the teacher is authorized. The system then verifies that the teacher has the appropriate role to create an exam. If successful, the teacher can select an exam and choose the question type. The "Select Exam" component returns the selected question type to the teacher, and they then proceed to add the question, providing its details. The "Question Bank" component saves the new question.

Figure 16 provides a step-by-step visual representation of the interactions between the teacher and the various system components, depicting the flow of information and actions involved in adding a question to an exam within the system, it highlights the key processes involved, including authentication, role verification, exam selection, question type selection, question adding, and saving the question to the question bank. This visual representation aids in understanding the system's behavior and identifying potential dependencies and interactions between different system components.

Figure 16. System-level sequence diagram for add question and exam by teacher

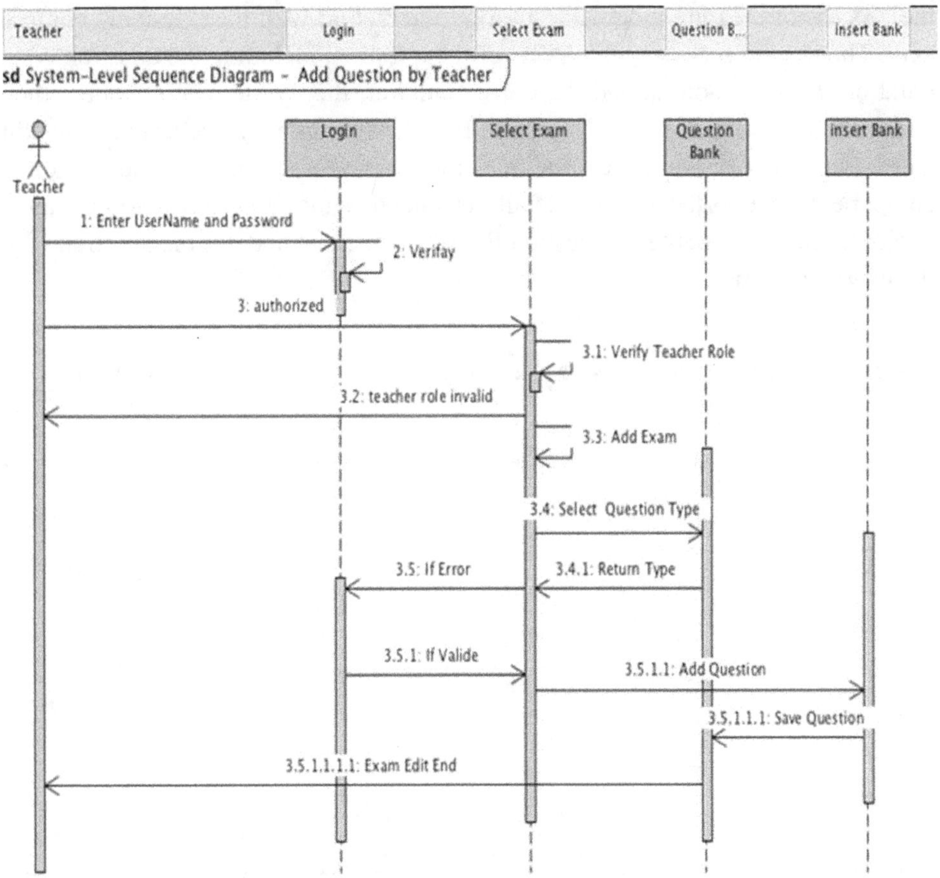

System-Level Sequence Diagram for Add or Edit Question by Teacher

Add or edit question by teacher sequence diagram provides a clear visual representation of how the teacher interacts with the system to add or edit questions, as it highlights the key steps involved, such as authentication, question creation or editing, and saving the question in the question bank. The diagram is a useful tool for understanding the flow of interactions and the various components involved in managing questions within the online exam system. As result it depicts the interaction between a teacher and the system when adding or editing a question within an online exam system. Figure 17 describes the sequence starts with the "User" (likely representing the teacher in this case) entering their username and password. This triggers a verification process by the "Teacher" component. If successful, the teacher can proceed to add a new question, then he "Teacher" component initiates the process of creating a new question by interacting with the "System" component. The teacher sets various aspects of the question, including the question type, duration, answer choices, and the correct answer, the "System" component then adds this new question to the "Question Bank.", alternatively, the teacher can edit an existing question, or selects the question they want to edit, and the "System" component retrieves the question details, teacher then modifies the question's name, choices, and duration, then "Question Bank" component updates the question with the new information.

Figure 17. System-level sequence diagram for add or edit question by teacher

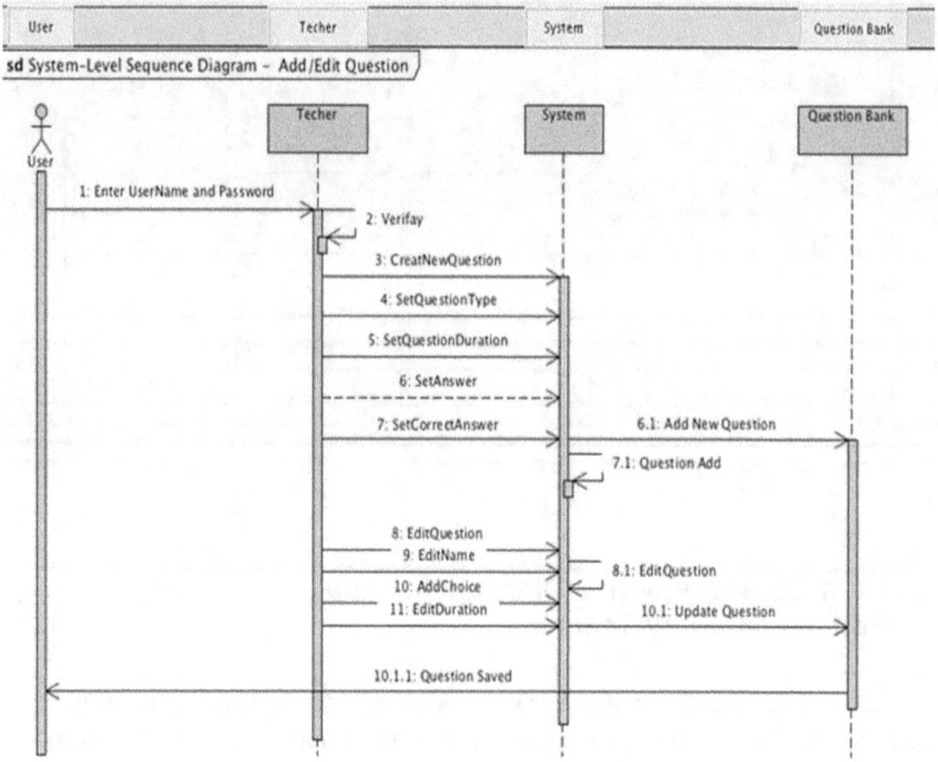

System-Level Sequence Diagram for Student How to Start Exam

Start Exam sequence diagram illustrates the interactions involved when a student starts an exam within an online exam system. It outlines the steps a student takes to begin an exam, focusing on the interactions with the system and the teacher.

Figure 18 begins with the "Student" actor, who initiates the process, and the teacher, which is represented by the "Teacher" lifeline, first logs into the system. The system then verifies the teacher's login credentials. The teacher then creates a group, adds a student, and registers for the exam. The student then logs into the system, selects the exam, answers the questions, and finally receives their exam result, it highlights the key processes involved in starting an exam, including teacher authentication, group creation, student registration, student authentication, exam selection, answering questions, and receiving results, it representation helps understand the system's behavior and identify potential dependencies and interactions between different system components.

Figure 18. System-level sequence diagram for student how to start exam

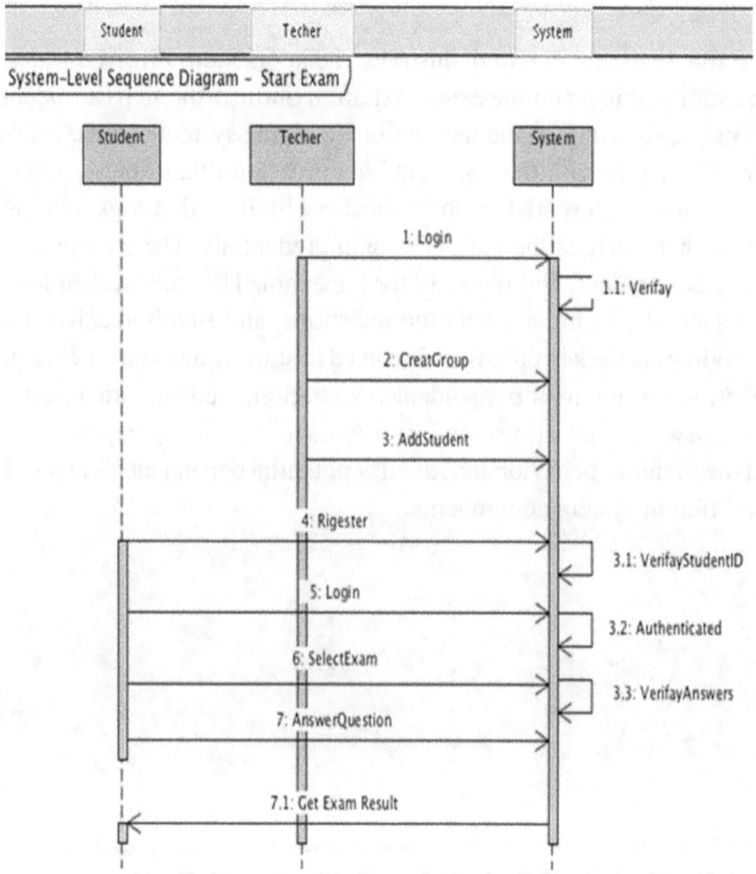

Activity Diagram

Activity diagrams, a type of behavioral diagram used in software engineering, are visual representations of workflows and processes within a system, it illustrates the flow of activities, actions, and decisions, often representing the interactions between different actors or components, activity diagrams are particularly useful for modeling business processes, workflow automation, and understanding complex procedures within a software system, it utilize symbols such as actions, decisions, forks, and joins to depict the sequence of steps, alternative paths, and parallel execution of activities. Activity diagrams are valuable tools for communication, analysis, and documentation, providing a clear and structured overview of how a system or process functions.

Activity Diagram: Completing the Exam

This activity diagram effectively illustrates the workflow involved in completing an exam, from the teacher's perspective of creating and grading the exam to the student's perspective of taking and submitting the exam. It highlights the distinct actions of each actor, the dependencies between them, and the overall process flow. This visual representation aids in understanding the system's behavior and identifying potential bottlenecks or areas for improvement in the exam completion workflow.

Figure 19 presents an activity diagram, a type of behavioral diagram used in software engineering, depicting the process of completing an exam in an online examination system, as it divided into two parallel flows, representing the actions of the teacher and the student. The teacher's flow begins with issuing an exam, which is then passed to the student, the student flow starts with selecting the exam, which is only available if they are part of the exam group, the student then proceeds to complete all the questions in the exam. Once finished, the student submits the exam. The teacher's flow continues as the deadline for the exam approaches. The teacher then grades the exam and saves the grade in the grade book. Finally, the teacher returns the graded work to the student. The student receives the graded work, and the process is complete.

Figure 19. Activity diagram: Completing the exam

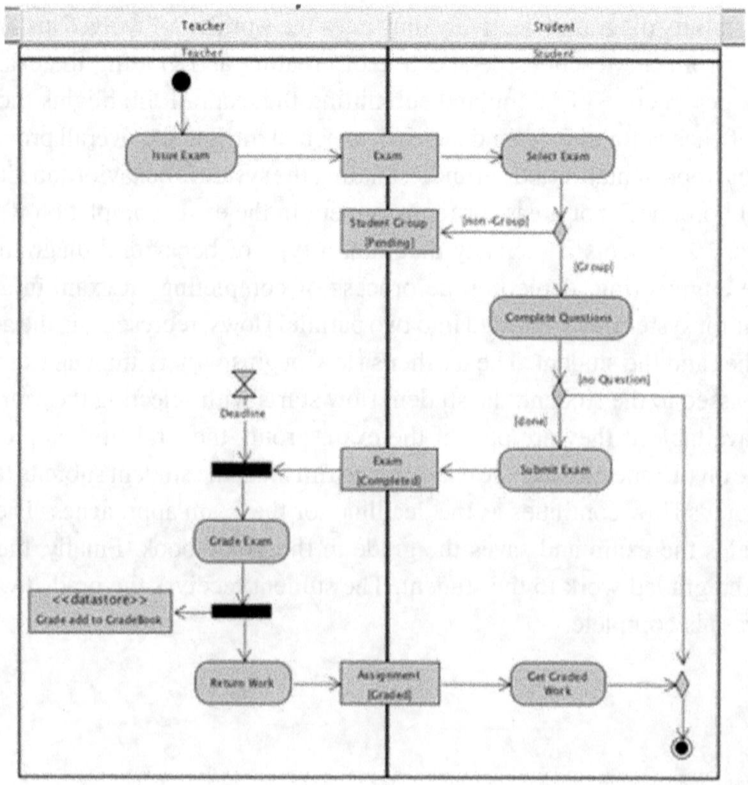

User Interface

Activity diagram effectively illustrates the workflow involved in completing an exam, from the teacher's perspective of creating and grading the exam to the student's perspective of taking and submitting the exam. It highlights the distinct actions of each actor, the dependencies between them, and the overall process flow, it aids in understanding the system's behavior and identifying potential bottlenecks or areas for improvement in the exam completion workflow. Figure 20 presents an activity diagram for the process of completing an exam it divided into two parallel flows, representing the actions of the teacher and the student. The teacher's flow begins with issuing an exam, which is then passed to the student, student flow starts with selecting the exam, which is only available if they are part of the exam group. The student then proceeds to complete all the questions in the exam. Once finished, the student submits the exam. The teacher's flow continues as the deadline for the exam approaches. The teacher then grades the exam and saves the grade in the grade book. Finally, the teacher returns the graded work to the student. The student receives the graded work, and the process is complete.

Figure 20. User main navigation path for the AEOS system

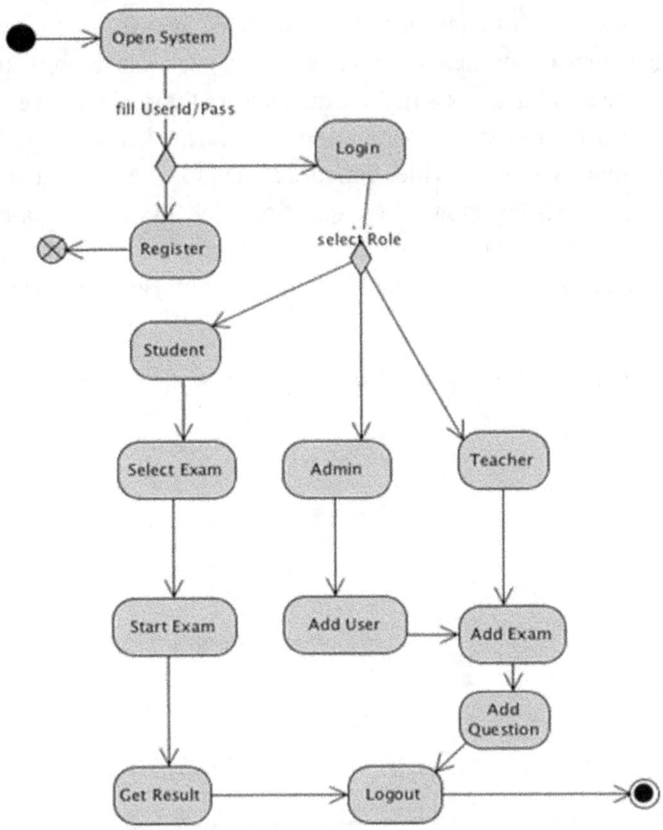

AOES Storyboard and Wireframes

The wireframes illustrate the overall layout and functionality of the AOES system, providing a visual representation of the user interface and how different roles interact with the system. The wireframes serve as a blueprint for the development of the final user interface, ensuring consistency and clarity in the system's design. Wireframes are skeletal outlines of a user interface, focusing on the layout and structure of a website or application, it is providing a visual representation of the arrangement of elements such as buttons, text fields, menus, and other interactive components without going into detailed design aspects like colors, fonts, or graphics. Figure 21 shows the storyboard and wireframes for AOES, which include login wireframe, administrator main page wireframe, teacher main page wireframe, student main wireframe, and exam main page wireframe. After successful login, each user role is directed to their respective main screen. The top part of the Figure shows a simplified hierarchical representation of the system's structure, a generic login screen branches out to three different main screens, one for each role: administrator, teacher, and student. The administrator's main screen allows them to manage users (adding, editing, deleting), create exams, manage the question bank, and view student results. The teacher's main screen provides functionalities for managing exams (creating, editing), adding questions, viewing the question bank, assigning exams to students, and managing the grade book. The student's main screen is focused on taking exams. Students can view their exam schedule, take the exams, and review their results.

Figure 21. Wireframes for student, teacher, and admin Wirefram

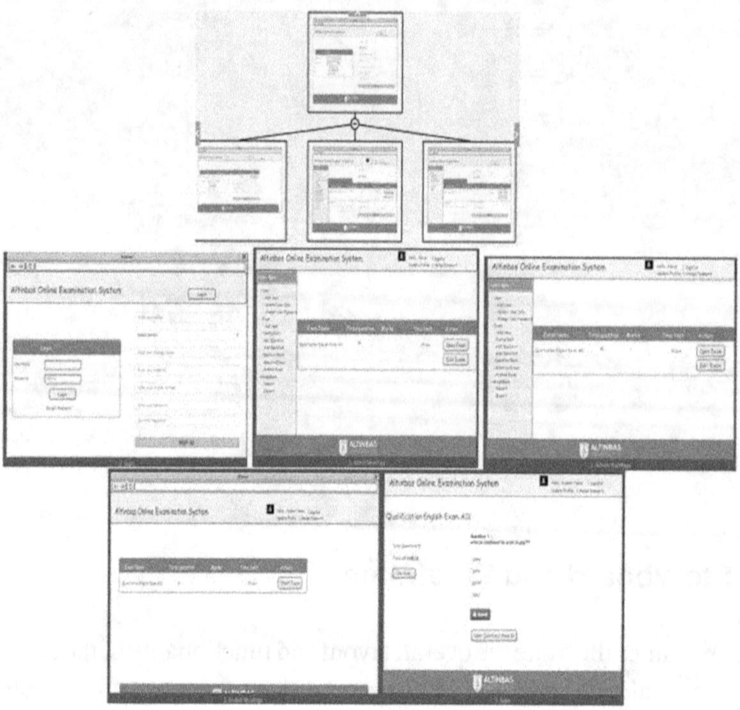

RESULT AND CONCLUSION

This chapter has taken a deep dive into the process of modeling the AEOS using various techniques from software engineering. We've explored how use cases and diagrams can be employed to define system requirements, visualize workflows, and communicate system behavior effectively, the techniques used in this chapter are widely applicable to various software engineering projects.

We started by identifying and defining the key actors involved in the AEOS system: the student, the teacher, and the administrator. Each actor has specific roles and responsibilities within the system, which are reflected in their respective use cases. We then employed various diagrams to model the system's functionality:

- Use Case Diagrams: It provided a high-level overview of the system's features and how different actors interact with it.
- Sequence Diagrams: Helped visualize the sequence of interactions between objects in the system, illustrating the flow of information and actions in specific scenarios like taking an exam or adding a question.
- Activity Diagrams: Modeled the workflow of specific processes, such as the steps involved in creating an exam, adding a question, or completing an exam.

By using these diagrams, we gained a deeper understanding of the system's behavior and dependencies between different components, these diagrams serve as essential tools for communication and collaboration between developers, stakeholders, and testers, ensuring a shared understanding of the system's functionality.

The Result show that as SW Engineering you can identified the core functionalities of AEOS, including user management, exam creation, question management, exam taking, and result generation, and defined the interaction patterns between different actors, such as how teachers create exams and add questions, how students take exams, and how administrators manage the system, All diagrams helped identify potential bottlenecks and areas for improvement in the exam process, such as the need for efficient question management or the requirement for clear communication between actors.

Use case diagrams can help gather comprehensive requirements by identifying actors, their goals, and the system's functionalities, sequence diagrams and activity diagrams can help visualize and understand complex interactions and workflows, leading to better-designed systems, use diagrams to effectively communicate system behavior and design decisions to stakeholders and team members, ensuring everyone is on the same page, these diagrams can serve as valuable documentation for testing and maintenance, aiding in understanding the system's structure and functionality over time.

By adopting these techniques, you can ensure that your software projects are well-defined, clearly communicated, and ultimately deliver a successful and robust solution.

CONCLUSION

The modeling techniques explored in this chapter are essential tools for software engineers, use cases and diagrams facilitate clear communication, guide design decisions, and help ensure a robust and user-friendly system, by applying these principles, you can create software systems that meet user needs effectively and efficiently.

REFERENCES

About the Unified Modeling Language Specification Version 2.0. (n.d.). Retrieved May 24, 2024, from https://www.omg.org/spec/UML/2.0/

Alter, S. (2001). Which Life Cycle—Work System, Information System, or Software? *Communications of the Association for Information Systems*, 7. Advance online publication. 10.17705/1CAIS.00717

Benfell, A. (2021). Modeling functional requirements using tacit knowledge: A design science research methodology informed approach. *Requirements Engineering*, 26(1), 25–42. Advance online publication. 10.1007/s00766-020-00330-4

Examino - Online Exam Maker. (n.d.). Retrieved December 9, 2022, from https://examino.finicode.com/ ISO/IEC

Jang, J. W., Verbeek, F., & Ravindran, B. (2021). Verification of Functional Correctness of Code Diversification Techniques. *Lecture Notes in Computer Science (Including Subseries Lecture Notes in Artificial Intelligence and Lecture Notes in Bioinformatics), 12673 LNCS*, 160–179. 10.1007/978-3-030-76384-8_11

Shankar, P., Morkos, B., Yadav, D., & Summers, J. D. (2020). Towards the formalization of non-functional requirements in conceptual design. *Research in Engineering Design*, 31(4), 449–469. Advance online publication. 10.1007/s00163-020-00345-6

USATestprep: K-12 standards-aligned practice tests. (n.d.). Retrieved December 9, 2022, from https://www.usatestprep.com/

Vishwas Kotwal, D., Rajendra Bhadke, S., Sanjay Gunjal, A., & Biswas, P. (2015). Online examination system. *International Research Journal of Engineering and Technology*. www.IJCSI.org

Chapter 4
Using Class Diagram and UML Component

ABSTRACT

The chapter utilizes a detailed class diagram to visually represent the system's components and their relationships, drawing upon the analysis conducted in Chapter 3. Furthermore, it explores dynamic models, including sequence diagrams, state diagrams, and activity diagrams, to capture the system's behavior and interactions over time. Finally, the chapter delves into AOES software architecture, employing UML package diagrams, component diagrams, and entity-relationship diagrams to provide a comprehensive view of the system's structure and organization.

INTRODUCTION

This chapter takes us deeper into the world of software design, building upon the foundation laid in the previous chapters. We've explored the basics of software engineering and the power of the Unified Modeling Language (UML) to capture system requirements. Now, it's time to translate those abstract concepts into concrete design elements, we'll explore how UML diagrams guide us in building a robust and functional software system, we dive into the intricate details of detailed design class diagrams, examining the attributes and methods of each class to accurately represent data structures and behavior. We'll delve into the nuanced world of relationships, including associations, compositions, and aggregations, and learn how to represent them using names and multiplicities. Moving beyond static structure, we'll explore the dynamic aspects of software systems using dynamic models, including sequence, state, and activity diagrams. These diagrams help visualize the flow of

DOI: 10.4018/978-1-6684-4844-1.ch004

interactions, state transitions, and activities within the system, offering a dynamic understanding of its behavior.

This chapter delves into the crucial phase of software design in the development of an online exam system (AOES). Building upon the analysis conducted in previous chapters, we translate abstract requirements into concrete design elements using the power of the Unified Modeling Language (UML).

We begin by exploring detailed design class diagrams, which visually represent the system's structure by outlining the classes, their attributes, methods, and the relationships between them. This detailed blueprint provides developers with essential information to implement the software effectively. Our practical example of AOES demonstrates how class diagrams capture the core elements of an online exam system: students, teachers, administrators, exams, questions, and results, showcasing how these elements interact and depend on each other. The diagram clearly illustrates the concepts of association, inheritance, aggregation, and composition, providing a comprehensive understanding of the system's static structure.

Next, we explore the dynamic aspects of the system using dynamic models, specifically sequence diagrams, activity diagrams, and state diagrams. These diagrams capture the system's behavior over time, highlighting the interactions between users and the system during key processes such as user registration, exam initiation, question management, and result retrieval. Each diagram offers a different perspective on the system's dynamics, showcasing the flow of actions, decisions, and state transitions.

The chapter then shifts focus to software architecture, employing UML package diagrams and component diagrams to illustrate the system's high-level structure. Package diagrams, as exemplified by the AOES system with subsystems, depict the organization of the system into logical modules or packages, promoting modularity, maintainability, and scalability. Component diagrams, on the other hand, illustrate the system's components and their interactions, highlighting the flow of information and control between different modules.

This chapter will also introduce you to essential UML diagrams for representing software architecture: package diagrams and component diagrams, these diagrams provide a high-level overview of the system's structure, clarifying the organization of components and their dependencies, finally, we'll bridge the gap between software design and database implementation by exploring Entity Relationship Diagrams (ERDs). ERDs map the data structures defined in our class diagrams to the corresponding database schema, creating a seamless connection between the software design and the underlying database structure. This chapter equips you with the tools and knowledge to transform abstract requirements into a detailed and well-structured software design. Mastering these techniques will enable you to create robust, efficient, and maintainable software systems.

By developing technology around the worlds, and most of trends in educational organization by using technology in education to support E-teaching method, adapting and using smart devises in one of the most important issues for applying Smart Teaching in such a university. This chapter proposes all the features and procedures to design the system for Our case used as a practical example AEOS for Abc University that will apply to English teacher. Design is the abstraction of a solution and view of patterns seen in the analysis phase to be a pattern in the design phase. In this chapter we will start to design AOES depending on the analysis report including all the requirements functional and non-functional requirements.

1. DETAILED DESIGN CLASS DIAGRAM

AOES is a web application that provides question to the student randomly should include three types of users (student, teacher, and administrator) as described in Figure 1, the main function of the system is to create the exam, display it to students to fill and submit and get the result, exam timer should be shown once the student starts Exam attempts, the system should satisfy the three sub-modules, next Figure summarize the user functions that they use, admin will have use all the functions that available to students and teacher, teacher can use student function after admin allow or disallow in user roll table, student can access the system and start his exam.

Figure 1. Context diagram of AOES

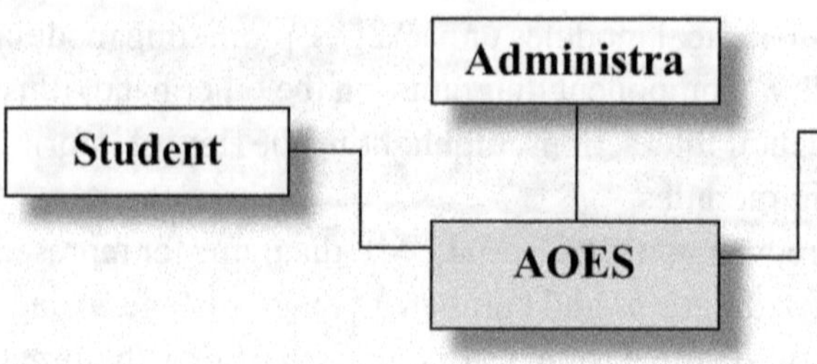

2. AOES SYSTEM DESIGN

Depending in analysis documentation the system can be viewed as five subsystems including the Manage user subsystem includes all users with the method we mentioned previously, The user subsystem will connect with Manage Account allowing the user to update their profile, the two subsystems will save data, and output to Database subsystem or local storage subsystem and all of the subsystem interact with Manage System as view operating system next figure describes the AEOS subsystem.

Figure 2. AEOS subsystem

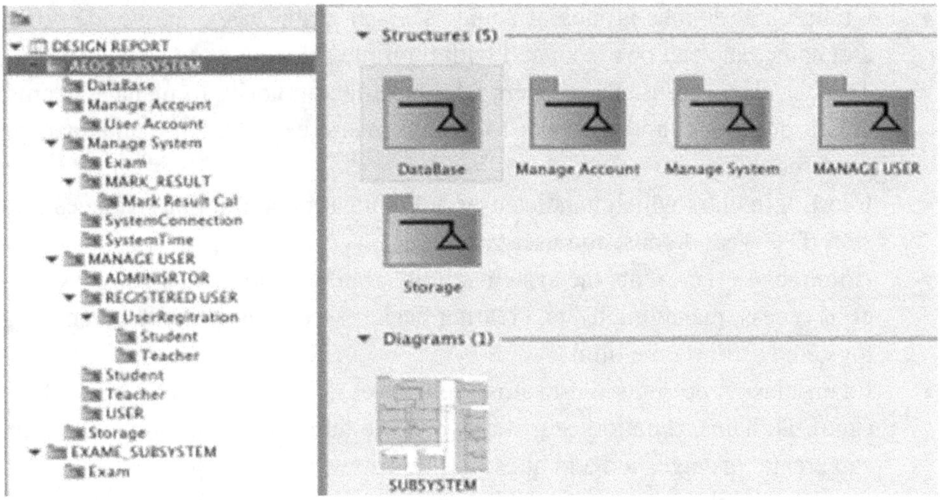

Class Diagram

The class diagram, often referred to as an "Object and Class Model" in software engineering, is a visual representation of the structure of a system. It depicts the classes that make up the system, their attributes (data they hold), and their methods (operations they perform), relationships between these classes are also illustrated, revealing how they interact and collaborate within the system. This diagram is a fundamental tool for understanding the system's architecture and how different components work together, it serves as a blueprint for developers, helping them to design, implement, and maintain the software effectively.

In essence, the class diagram provides a high-level view of the system's static structure, capturing the key elements and their relationships. It's like a blueprint for a building, showing the rooms, their layout, and how they connect. By understanding the class diagram, developers can gain a clear picture of the system's design and how it will function.

The class diagram practical example used in this book for the AOES package illustrates in Figure 3 the relationships between various classes using UML notation to represent the relationships between the below classes and their interactions:

- Student class holds information about individual students, including their name, ID, group, status, password, and other relevant details. Students can register, take exams, view results, and reset their passwords.
- Group class defines groups of students. Each group has a unique identifier, and name, and can be associated with multiple students.
- Teacher class represents teachers who have the authority to manage exams, create questions, add students to groups, view exam results, and manage grade books.
- UserLogIn class which handles user authentication and login, manages session IDs, event details, and user types.
- Admin class represents the system administrators, who have the highest level of access, managing users, creating backups, recording results, managing logs, and setting time limits.
- Exam class represents the exams themselves, holding information such as exam ID, name, duration, start time, and the number of questions. Teachers can create, manage, and add questions to exams.
- Question Bank class stores and manages questions, it includes information about the question ID, name, type, choices, duration, and correct answers.
- Student Attempt class captures a student's attempt on an exam, including their response time, question ID, selected answer, and student ID.
- Results class represents the results of an exam, including the student's ID, exam ID, marks, and other relevant information.

Figure 3 illustrates the main relationships between entities involved in managing exams (students, teachers, and the overall system). This diagram is a valuable tool for understanding the system's architecture and how different components interact as shows several key relationships between these classes:

1. Association

The 'Student' class is associated with the 'Group' class through the “Assign” relationship, indicating that students belong to groups. Similarly, teachers manage students and exams. This relationship implies a connection between two classes, where one class is aware of the other but doesn't contain or own it.

2. Inheritance

The 'UserLogIn' class is inherited by both the 'Student' and 'Teacher' classes, indicating that both types of users need to log in. Inheritance represents an “is-a” relationship where a subclass inherits the properties and methods of its parent class.

3. Aggregation

The 'Exam' class aggregates the 'QuestionBank' class through the “Add” relationship, implying that exams consist of questions from the bank. Also, the 'Teacher' class aggregates the 'StudentAttempt' class through the “Submit” relationship, suggesting that teachers can view student attempts. Aggregation represents a “has-a” relationship where an object contains another object but doesn't own it.

4. Composition

The 'Exam' class is composed of 'Results' through a “1..*” relationship. This indicates that each exam can have one or more results. Composition is a stronger form of “has-a,” indicating that the contained object cannot exist without the containing object.

Figure 3. Class diagram for AOES package

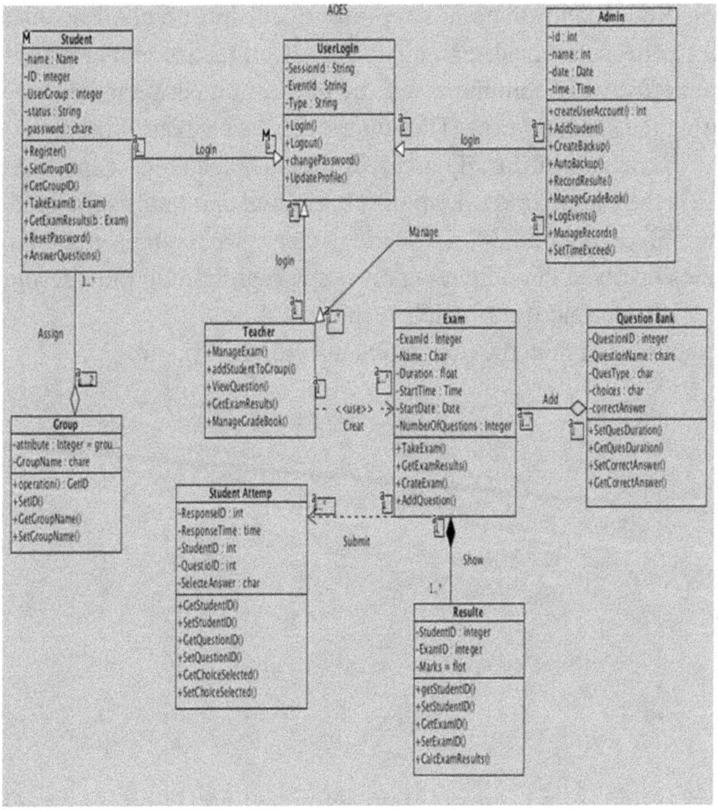

AOES User Main Module

To illustrate the process using an activity diagram by UML to visually depict the flow of activities involved, splitting the process into Teacher and Student, showcasing the interactions between the system and the users, below is how the diagram uses UML elements to illustrate the process:

Figure 4 is divided into two verticals representing the three main actors involved – "Teacher", "Student" and "admin", it separates the actions performed by each. First, we start by initial black which fills the circle at the top of the Teacher to represent the start of the process. Then Action Nodes which are represented by rounded rectangles, depict individual actions in the process. Examples include "Select Exam," "Complete Questions," "Submit Exam," and "Grade Exam". Control Flow as arrows which indicate the sequence of actions, for example, after the teacher issues the exam, the process moves to the students, where the student selects the exam.

Decision Nodes, and diamonds represent decision points. For instance, after the "Select Exam" action, a decision node checks if the student belongs to a group. The diagram effectively communicates the interaction between the student and the system during the exam process. The student selects an exam, completes questions, submits the exam, and ultimately receives their graded work. Simultaneously, the teacher issues the exam, grades it upon submission, and returns the graded work to the student. This activity diagram provides a clear visual representation of the steps and decisions involved in completing an exam, highlighting the interplay between the student's actions and the system's responses.

main function and how the user interacts with the system.

Figure 4. User with main module

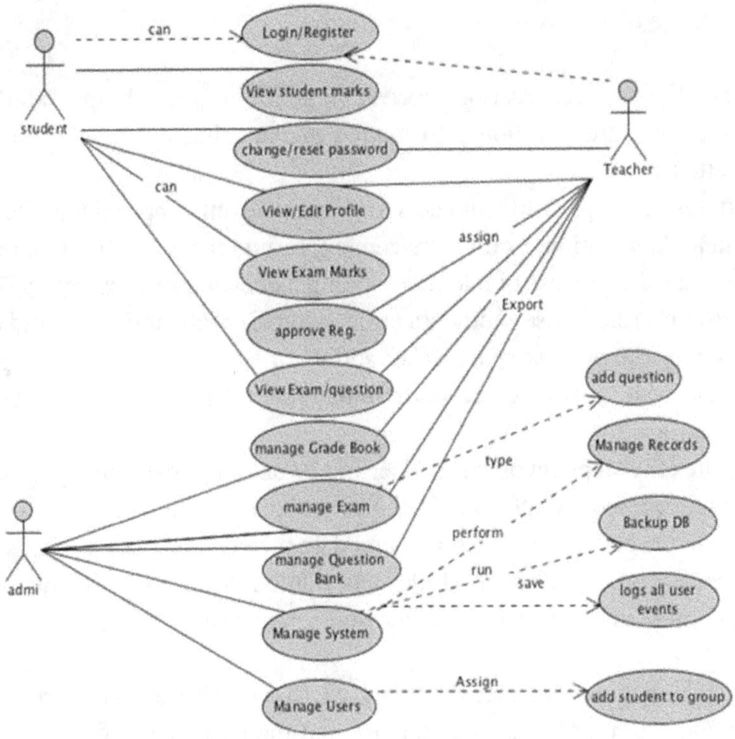

3. DYNAMIC MODELS

Dynamic models are a crucial aspect of software engineering, providing a visual representation of a system's behavior over time, they capture the interactions between components, the sequence of events, and the data flow, offering valuable insights into how a system operates dynamically. Among the dynamic modeling techniques, the sequence diagram stands out as a particularly effective tool for illustrating the flow of messages between objects in a specific scenario.

Sequence diagrams are a powerful tool in software engineering, we provide practical Example to a visual representation of the interactions between different actors and system components over time, sequence diagrams can effectively illustrate the flow of events for various key processes, such as user registration, exam initiation, question management, and result retrieval. In this section a series of sequence diagrams, each focusing on a specific aspect of the online exam system. These diagrams offer valuable insights into the system's dynamic behavior, showcasing the intricate interplay between users (teachers, students, administrators) and the system itself, list of Sequence diagrams we provide as follow:

1. outlines the user registration process, highlighting the administrator's role in receiving new registration information and making decisions based on the submitted data.
2. illustrates the steps a student takes to begin an exam, emphasizing the system's role in authentication, exam management, result retrieval, and mark recording. Showcases the process of a teacher adding a question to an exam, highlighting the system's role in user authentication, authorization, and data validation.
3. Focuses on the process of a teacher adding or editing a question, emphasizing the sequential steps involved in defining and updating question details within the system.
4. illustrates the steps involved in inserting or updating data, likely related to an exam management platform, highlighting the interaction between the teacher, the login process, exam selection, and data management components.
5. outlines the process a student follows to start an exam, emphasizing the collaborative process between the teacher, student, and the system.

By analyzing these sequence diagrams, we gain a comprehensive understanding of the online exam system's functionality and the dynamic interactions between its various components. This understanding is crucial for effective system design, development, and maintenance.

3.1 Sequence Diagram

To demonstrate the power of sequence diagrams, let's consider the common process for AOES example, a sequence diagram for user registration would typically involve actors like the user, the registration form, the system's backend logic, and the database, the diagram would clearly show the user interacting with the registration form, filling in required information, and submitting the form. The system would then validate the data and, if valid, send a message to the database to create a new user record. Upon successful creation, the database confirms back to the system, which then sends a success message to the registration form. Finally, the user receives a confirmation message. This sequence diagram visually captures the entire registration process, step by step, highlighting the interactions between the user and the system. Sequence diagrams provide several benefits in software development. They visually depict the order of events and message exchanges, facilitating a clearer understanding of system behavior. They enhance communication between developers, designers, and stakeholders by providing a shared understanding of the system's dynamic workings. They serve as essential documentation for the system's functionalities and interactions. The applications of sequence diagrams extend beyond user registration, encompassing a wide range of dynamic processes, such as login and authentication, order processing, payment transactions, and data retrieval and updates. Ultimately, sequence diagrams are a powerful tool for modeling the dynamic aspects of software systems, offering a clear, concise, and visual representation of object interactions, improving the overall understanding, communication, and documentation of system behavior.

Figure 5 outlines the user registration process within a system, it depicts the interactions between various actors and system components: the administrator, a new registration entity, the registration process, a database ("Receive master DB"), an exam master, a teacher master, and a system master. The process begins with the administrator receiving information about a faculty course subject. If this subject is new, the registration process initiates of "Insert" operations, likely indicating the creation of new entries in the system. These insertions occur sequentially, first in the registration process itself, then in the database, followed by the exam master, teacher master, and finally, the system master, after all the insertions are completed, the administrator receives an "Accept/reject" prompt, suggesting a point of decision based on the inserted data. Once the administrator makes a decision, the process flow ends. The sequence diagram effectively highlights the linear flow of actions and information exchange during user registration. It emphasizes the role of the administrator in initiating the process and making decisions based on the newly registered data, the sequential "Insert" operations suggest a step-by-step process of adding information into different components of the system, ensuring consistency and proper data management during user registration.

Figure 5. User registration sequence diagram process

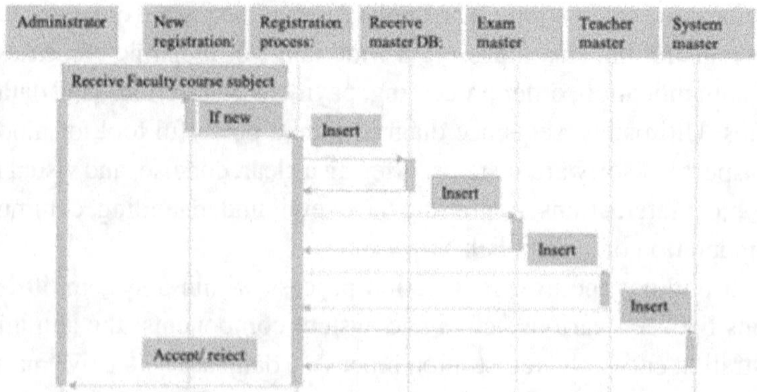

The sequence diagram effectively outlines the flow of events, highlighting the interactions between the student and the system throughout the exam process. It emphasizes the system's role in authentication, exam management (starting, timing, and ending), result retrieval, and the recording of marks. It also shows the alternative flow in case a student doesn't complete the exam within the allotted time, ensuring a robust system design.

A system-level sequence diagram as outline in Figure 6, illustrates the steps involved when a student begins an exam in an online exam system, it details the interactions between the student and the system from login to the recording of the exam score. The diagram begins with the student entering their username and password for login, the system then verifies the credentials, if valid, the student is authorized to proceed, the system checks the student's group affiliation. If validated, the student can then begin answering exam questions. "Start Exam" component represents the exam commencement, while "End Exam" signifies its conclusion. During the exam, a timer ("Exam Time") tracks the allotted time. When the time expires, or the student finishes all questions, the student submits the exam ("Submit Exam"). Following submission, the system retrieves the exam results ("Get Exam Result") and records the student's marks ("Set Marks") in the grade book ("Grade Book"). The diagram also illustrates an alternative flow: If the student doesn't finish the exam within the allotted time ("Finish Exam"), the system will still process the exam and record the marks based on the answered questions.

Figure 6. System-level sequence diagram for student when starting the exam

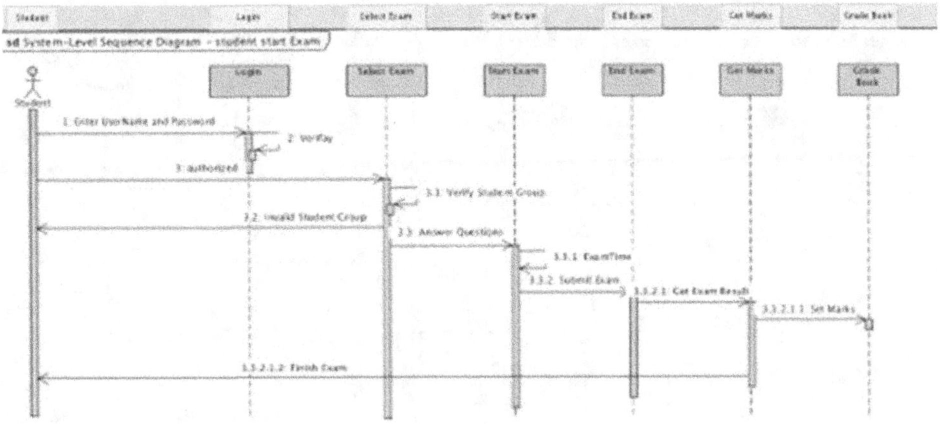

Add Question to Exam Sequence diagram effectively captures the step-by-step process of a teacher adding a question, including login, role verification, exam selection, question type selection, and finally, adding and saving the question. It demonstrates the system's role in user authentication, authorization, data validation, and interaction with the question bank. The diagram also illustrates the alternative flow in case of an error, showcasing a robust system design that handles potential issues.

Figure 7 is a system-level sequence diagram illustrating the process of adding a question to an exam by a teacher in an online exam system. The diagram outlines the steps involved, highlighting the interactions between the teacher and the system, this process starts with the teacher logging in by providing their username and password. The system then verifies the credentials. If valid, the teacher is authorized to proceed. The system verifies the teacher's role to ensure they have the necessary privileges to add questions. If the teacher's role is invalid, the process ends.

After role verification, the teacher can add an exam ("Add Exam") and then select the type of question they want to add ("Select Question Type"). The "Question Bank" component likely represents the database or repository where questions are stored. The system returns the selected question type ("Return Type") to the teacher. The diagram then presents a decision point ("If Error"). If an error occurs during the process, the flow ends. However, if the data is valid ("If Valid"), the teacher can add the question ("Add Question") and save it ("Save Question") to the question bank ("Insert Bank"). The process concludes with the end of the exam editing session ("Exam Edit End").

Figure 7. System-level sequence diagram for add question and exam by teacher

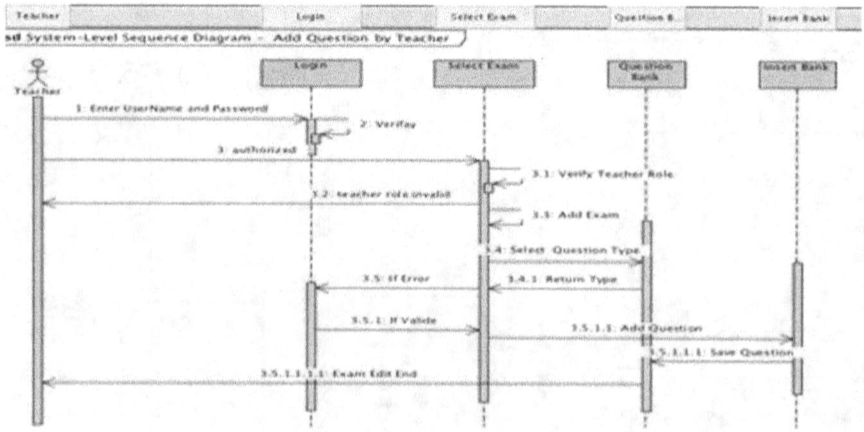

Add or edit question by teacher sequence diagram effectively captures the step-by-step process of adding or editing a question, demonstrating the system's role in user authentication, handling question details, and interacting with the question bank. It emphasizes the sequential nature of these operations, highlighting the specific order in which information is exchanged between the teacher and the system. This diagram serves as a clear visual guide for understanding the workflow of question management within the online exam system, a system-level sequence diagram as outlined in Figure 8, showcases the process of a teacher adding or editing a question within an online exam system. It outlines the interactions between the teacher, the system, and the question bank.

The process kicks off with the teacher providing their username and password to log into the system, the system then verifies the credentials. After successful authentication, the teacher can initiate the process of creating a new question ("CreateNewQuestion"), the sequence diagram then depicts the steps involved in defining the question: setting the question type ("SetQuestionType"), duration ("SetQuestionDuration"), providing the answer ("SetAnswer"), and setting the correct answer ("SetCorrectAnswer"). The system then adds this new question ("Add New Question") to the question bank.

The diagram further outlines the steps for editing an existing question, the teacher initiates the editing process ("EditQuestion"), then proceeds to modify aspects of the question, such as its name ("EditName"), choices ("AddChoice"), and duration ("EditDuration"). The system then updates the question in the question bank ("Update Question"). Finally, the system confirms that the question has been successfully saved ("Question Saved").

Figure 8. System-level sequence diagram for add or edit question by teacher

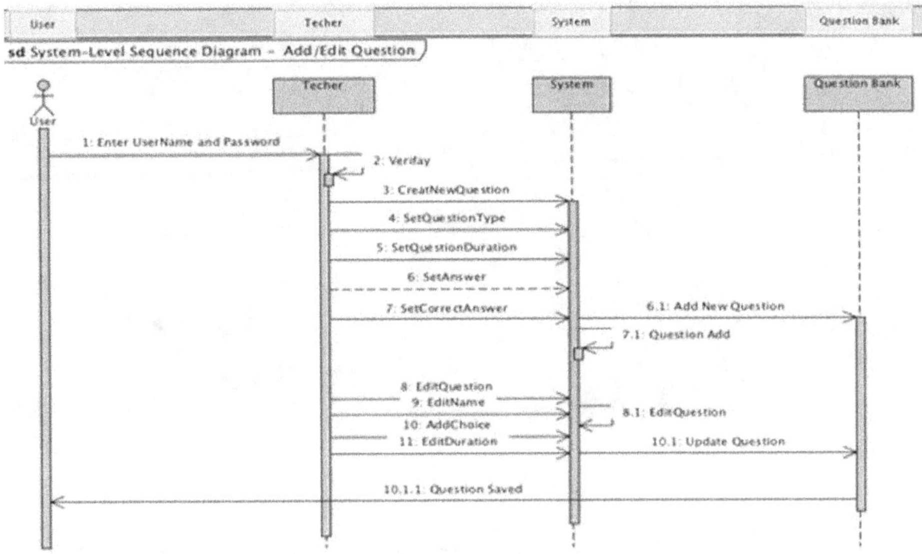

Insert or update operation sequence diagram, it emphasizes the role of the teacher in initiating the process, the system's role in authentication and data handling, and the multiple decision points that ensure data integrity and user confirmation throughout the process. Figure 9, a sequence diagram, illustrates the process of inserting or updating data within a system, likely related to an exam management platform. The diagram involves interactions between a teacher, a login process, an exam selection process, a subject question entity, a process for inserting questions, and a "Question Add to" entity.

The process begins with the teacher entering their username and password for login. The system then verifies the credentials. Following successful verification, the "Select Exam" component requests an exam. Simultaneously, the teacher may engage in "Exam edit" activities. Afterward, a decision point emerges where the system prompts for acceptance ("Accept/"). If accepted ("If Accept"), the "Select Exam" component proceeds to "Add question." This action triggers the "Subject Question" component to prepare the question for insertion ("Insert question"), which is then sent to the "Question Add to" component. The "Question Add to" component then "Store question," indicating the data is saved or updated. The final step involves another decision point. Both the "Insert question" and "Question Add to" components prompt for an "Accept/reject" decision. This likely represents a confirmation or final approval step before the inserted or updated data becomes permanent within the system.

Figure 9. Sequence diagram for insert or update operation by administrator

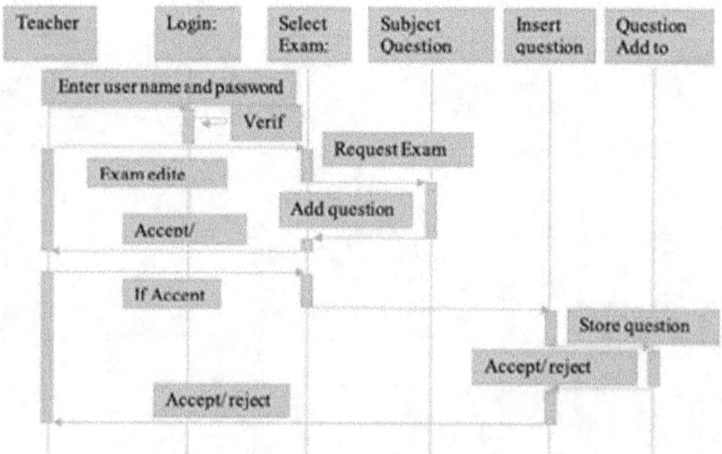

System-level sequence diagram include teacher, student, and the system for starting an exam start by the teacher's role to setting up the environment, then the student's actions in taking the exam, and the system's crucial functions of authentication, verification, and result retrieval, it emphasizes the importance of security and data integrity throughout the exam process, ensuring a reliable and secure exam environment.

Figure 10, a system-level sequence diagram, outlines the steps a student takes to start an exam within an online exam system. The diagram highlights the interactions between the student, teacher, and the system. The process begins with the teacher performing initial setup actions. This includes logging in (“Login”), creating a group (“CreatGroup”), and adding the student to that group (“AddStudent”). Once these steps are completed, the student can begin the exam process. The student first registers with the system (“Register”), then logs in using their credentials (“Login”). After successful login, the student selects an exam (“SelectExam”). At this point, the system interacts to verify the student's ID (“VerifyStudentID”) and authenticate their access (“Authenticated”). The student then proceeds to answer the exam questions (“AnswerQuestion”). The system then verifies the student's answers (“VerifyAnswers”). Finally, the student can retrieve their exam results (“Get Exam Result”).

Figure 10. System-level sequence diagram for student how to start exam

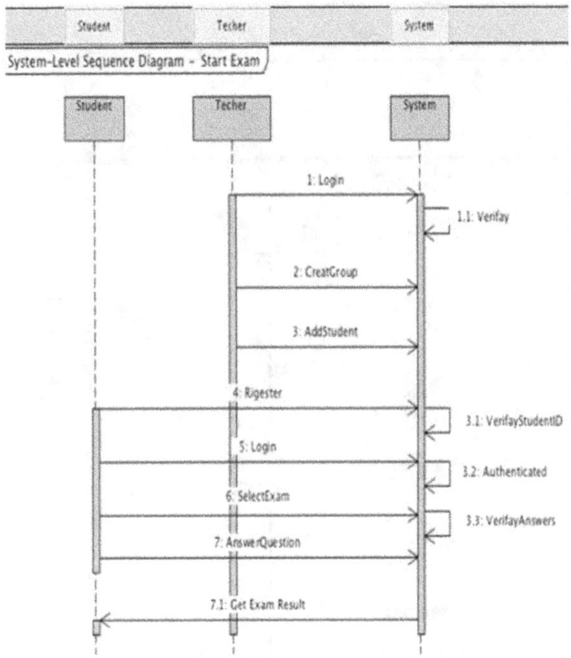

3.2 Activity Diagram

As we mentioned previously to visually represent the flow of actions within a system and to provide a clear overview of the steps involved in a particular process we utilize a variety of symbols, including action nodes, decision nodes, and control flows, we illustrate the sequence of activities and decision points within a workflow as outline in Figure 11 the activity diagram include all process of completing an exam in an online exam system, clearly separating the actions of the "Teacher" and the "Student" within their respective, the process begins with the teacher issuing the exam then student selects the exam and, depending on their group affiliation, proceeds to complete the questions. The diagram depicts decision points, such as checking if there are any remaining questions or if the exam deadline has been reached. Once the student submits the exam, the teacher grades it, adds the grade to the grade and finally returns the graded work to the student. The activity diagram effectively highlights the interplay between the actions of the teacher and the student throughout the exam completion process, providing a comprehensive visual representation of the workflow.

Figure 11. Activity diagram: Completing the exam

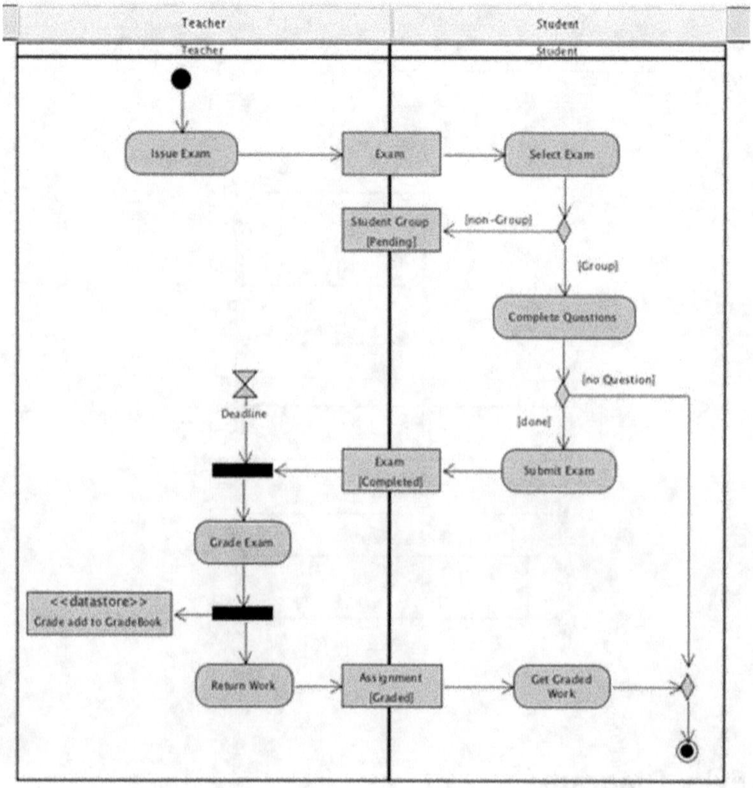

3.3 State Diagram

In software engineering, state diagrams provide a visual representation of the different states a system can be in and how it transitions between these states based on specific events or conditions. They depict the lifecycle of an object, showcasing the various states it can occupy and the triggers that cause it to shift between those states.

Figure 12 is a state diagram for AOES users with main navigation and main path steps, It illustrates the main navigation path for users within the AOES system. Starting with the "Open System" state, users can either choose to register or log in. If they choose to register, they will be directed to the "Register" state. If they choose to log in, they will need to provide their user ID and password, and then proceed to the "Login" state. After successfully logging in, the user selects their role - "Student," "Admin," or "Teacher." Each role has its path: students can "Select Exam," "Start Exam," and "Get Result"; admins can "Add User"; and teachers can "Add Exam" and "Add Question." All users, regardless of role, can "Log out" of the system, leading to the final state, represented by the filled circle. This state diagram effectively outlines the different states a user can be in within the AOES system, showcasing the paths and actions available based on their role and choices.

Figure 12. User main navigation path for the AEOS system

To describe the activity in detail we need some examples to visually depict the flow of actions and decisions within a process, offering a clear step-by-step representation of a system's workflow. However, when you need to visualize the different states an object or component can be in and how it transitions between those states, a state diagram is more appropriate. Figure 13 showcases the process an AOES teacher follows to create a question for an exam, the diagram starts with the teacher, who can then "Insert question," leading to the "Insert Questions" state. The next stage involves interacting with a "Question Type master," likely a component responsible for managing question types. From there, the diagram transitions to the "Question answers" state, where the teacher provides the answers to the question. Finally, the process reaches its end state, represented by a filled circle. This state diagram effectively depicts the different states involved in creating a question, showing the sequential flow from the teacher's action to the finalization of the question with its answers.

Figure 13. State diagram for AOES teacher for create question to exam

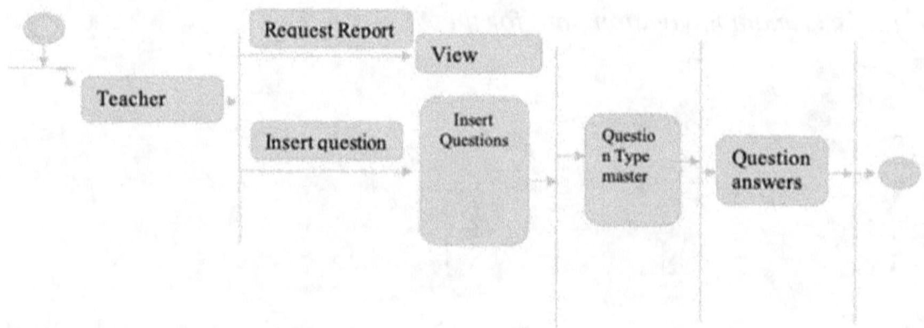

Figure 14, a state diagram that provides a clear understanding of the object's lifecycle and the triggers that cause it to change states, outlines the steps an AOES student takes when attempting an exam. The student starts in the "Student" state, and after registering ("Register") can view information ("View"). The student then progresses to "Start Exam," followed by "View Question", then branches into three separate states: "Exam Time," where the exam timer is active, "Exam Result," where the student can view their results, and "Control Logs," which likely represents a state for monitoring exam activity. The diagram concludes with a final state represented by a filled circle, indicating the end of the exam attempt process.

Figure 14. State diagram for AOES student to attempt to exam

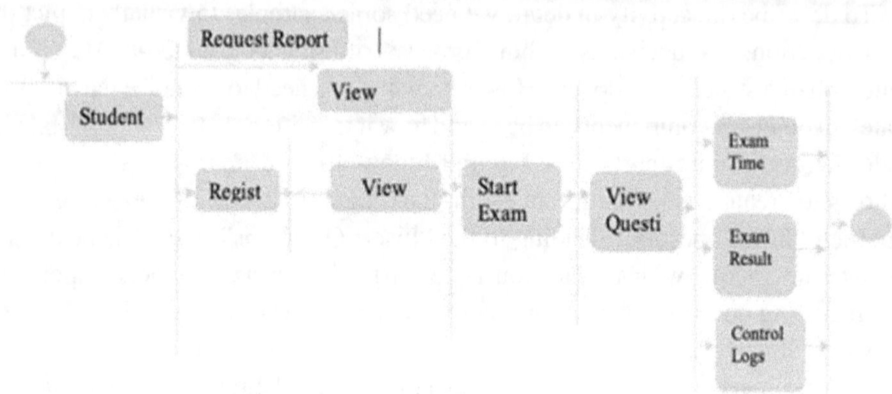

4. SOFTWARE ARCHITECTURE

Software architecture is the high-level structure of a software system, encompassing its components, its relationships, and the principles guiding its design and evolution. It provides a blueprint for the system, outlining how different elements work together to achieve the desired functionality as shown in Figure 15 which provides a compelling illustration of software architecture in action, it reveals a well-defined architectural structure, dividing the system into distinct subsystems or packages. The main package, "AOES Subsystem," encapsulates the core functionalities of the system, further dividing them into "Manage User," "Manage Account," and "Manage System." This modular approach promotes better organization, maintainability, and scalability.

By depicting the system's components, their relationships, and the guiding principles of modularity and separation of concerns, it provides a clear, high-level blueprint for understanding the system's structure, organization, and interdependencies, serving as a valuable tool for developers, designers, and stakeholders alike.

4.1. AOES UML Package Diagram Package

Figure 15 presents a UML Package Diagram for AOES, it provides a high-level view of the system's architecture, showcasing the key subsystems and their dependencies, as it is organized into several packages, each representing a distinct area of functionality.

The main package, "AOES Subsystem," contains sub-packages for managing users, exams, and the system itself. The "Manage User" package includes functionalities like managing users, registering users, and updating profiles. It further segregates user roles into "Administrator," "Teacher," "User," and "Student," indicating different levels of access and privileges. The "Manage System" package encompasses functionalities related to exams, mark calculation, system time, and data connections. It includes packages for managing exams, calculating marks, retrieving system time, and establishing system connections. The diagram also illustrates dependencies between these packages, showing how they interact with each other. For instance, the "CreateExam" function interacts with the "Exam" package within "Manage System." Similarly, the "save data" function interacts with both "Storage" and "Database" packages.

The package diagram effectively visualizes the system's structure, highlighting the modularity of its design and the relationships between its different components. This diagram is a valuable tool for understanding the system's architecture and how different functionalities are grouped and interconnected.

Figure 15. AOES system with subsystem

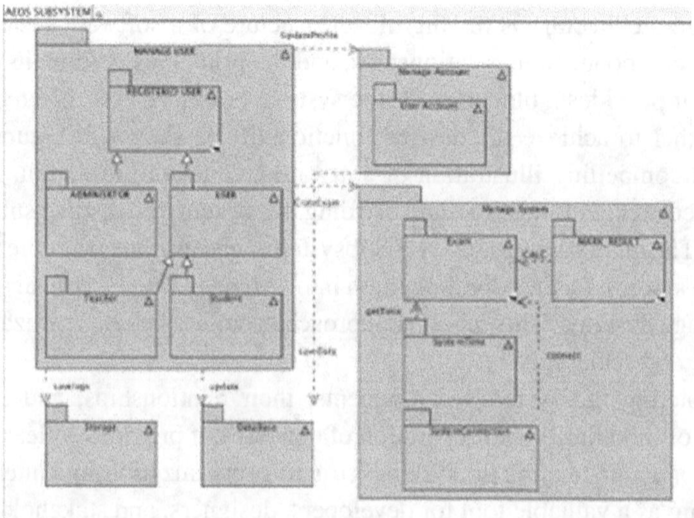

Each element represents a distinct functionality related to exams, and their relationships demonstrate how they collaborate to support the overall process of exam creation and management. This modularity promotes code reusability, maintainability, and easier troubleshooting, as changes made to one element are less likely to impact others.

Figure 16 focuses on the "Exam" subsystem within a larger online exam system. It highlights the key elements and their interactions within this subsystem, providing a detailed view of how exam-related functionalities are structured and organized, it utilizes a package structure, visually represented as a folder labeled "Exam." Within this package are several elements, each represented as a rectangle, that encapsulate specific functionalities. These elements include "AddExam," "AddQuestion," "SetTime," "CorrectAnswer," "QuestionOptions," and "Teacher."

The relationships between these elements are indicated by dashed lines with arrowheads. The "Teacher" element has associations with all other elements within the "Exam" package, suggesting that teachers are responsible for creating exams, adding questions, setting time limits, defining correct answers, and managing question options.

Figure 16. Exam sub-system

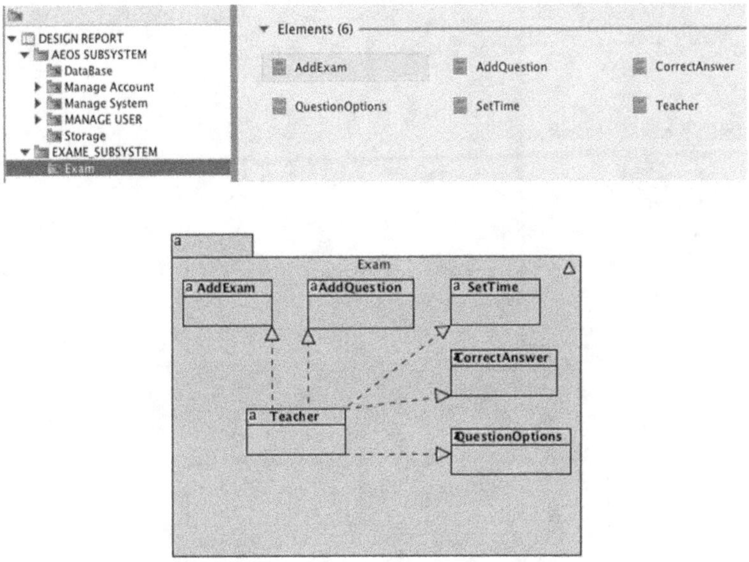

Figure 17 illustrates the "Exam Result Sub System" within a larger online exam system. This diagram focuses on the elements involved in calculating and presenting exam results. The diagram uses a package structure, represented by a folder labeled "Mark Result Cal," which contains the core components of this subsystem. The key elements within this package include "Exam Result," "Correct Answer," "System/user," and "Question Mark."

The "correct Answer" element is central to the process. It has an association with "Question Mark," suggesting that each question has a designated correct answer. The connection between "System/user" and "Question Mark," labeled "selected choice," indicates that the system tracks the user's chosen answer for each question. A dashed line connecting "selected choice" and "correct choice" suggests a comparison process to determine if the user's answer matches the correct answer. The "calc" association between "System/user" and "Exam Result" indicates the system's role in calculating the overall exam result based on the comparison between selected and correct answers.

This diagram effectively showcases the process of exam result calculation within the subsystem. It highlights the key elements involved: the correct answer, the user's selected answer, the question's mark value, and the final exam result. The diagram's emphasis on the comparison between selected and correct answers underscores the accuracy and integrity of the system's result calculation process.

Figure 17. Exam result sub system

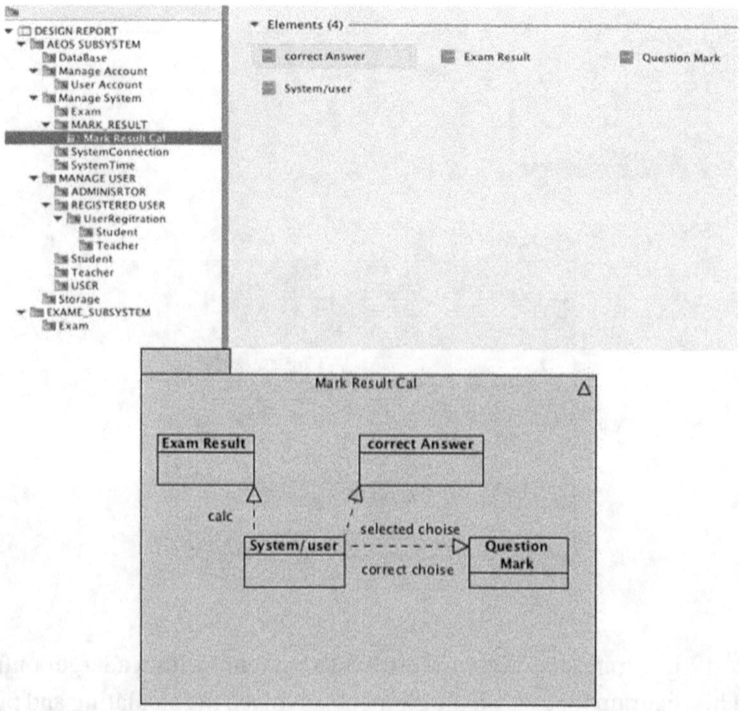

UML Component Diagram

Component diagram depicts the high-level architecture of an online exam system, showcasing the key components and their interaction, it employs UML notation to represent components, interfaces (represented as circles), and dependencies (lines connecting components).

component diagram effectively illustrates the system's modular architecture, as each component encapsulates a specific functionality, promoting maintainability and reusability, the dependencies between components clearly show the flow of information and control within the system. This diagram provides a valuable overview of the system's architecture, facilitating communication and understanding among developers and stakeholders.

Figure 18 outline the process which starts with the "User" who can either "Log-in" to an existing account or "Register" a new one. Both "Login" and "Register" components interact with the "DataStore" to verify credentials or store new user data. The “DataStore” is likely a database, responsible for persistent storage of all system data.

The core functionality of the system revolves around the “Exam” component, “User” interacts with “Exam” to take exams. “Exam” uses the “GetQuestion” interface to retrieve questions from the “QuestionBank.” Upon completion of an exam (“FishExam” interface), the “Exam” component communicates with the “ExamResult” component to calculate and present the exam results.

Figure 18. UML component diagram

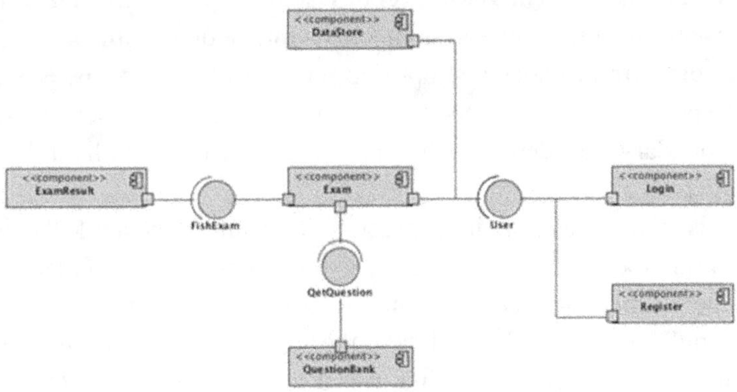

5. ENTITY RELATIONSHIP DIAGRAM

In this section, we apply data flow diagram to show the flow of data by user, level 0 describes the system user for AOES with the data, then we describe in details for system user level 2 and level 2 for users include student, teacher and administrator, depending on Data Flow Diagram (DFD) we use to identify ERD as describe below. The ERD is used at the design stage to identify all elements of the future system and their relationships and dependencies. Later we use the ERD stage as the future structure and functionality depend on it (Ľachová & Trebuňa, 2019).

Within the realm of DFDs, the context diagram is the most abstract level (Level 0), it focuses on portraying the system as a black box, hiding its internal complexities and emphasizing the primary data flows between the system and external entities. This high-level perspective is crucial for understanding the system's context, its role within a larger environment, and its primary interactions with the outside world (Zhang et al., 2018).

5.1 A Data Flow Diagram

DFD maps out the flow of information for any process or system, Next example describes the DFD for AOES for context diagram, first level, and second level, DFD is very important to present how AOES will interact with module and users in the system, DFD used as an excellent tool for summarizing and organizing all system detailed such as boundaries, processes, and data entities (Davis, 2019).

Context Diagram

Context diagrams are the highest level of Data Flow Diagrams (DFDs), providing a basic overview of the entire system and its interactions with external entities. This context diagram provides a simplified, high-level view of the System, as it illustrates the primary data flows between the system and its key users. It establishes the system boundary and identifies the main external entities without delving into the internal details of the system.

Figure 19 is represented as a single process (the rounded rectangle labeled "level 0"). This central process interacts with two external entities: the "Teacher" and the "Admin". The teacher interacts with the system by providing login information ("Validate login"), the system then validates the teacher's credentials. Admin. Admin interacts with the system by submitting user detail requests ("User Details Req.") and receiving validation of their username and password ("Username, Pwd"). This diagram sets the stage for more detailed DFDs (level 1, level 2, etc.) that will break down the "Online Exam System" process into its constituent sub-processes, revealing the internal data flows and logic of the system.

Figure 19. Context diagram for AOES

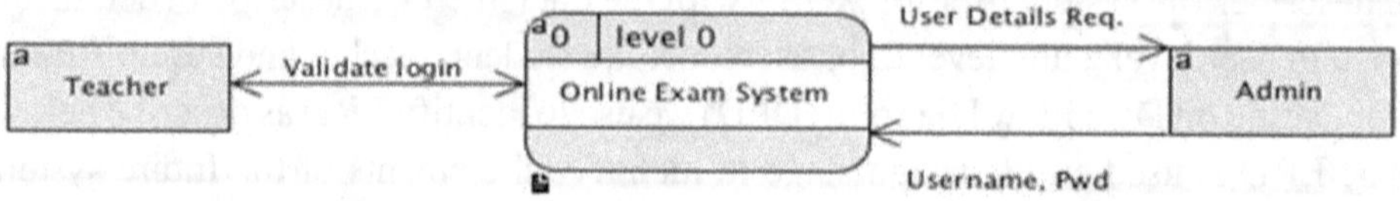

First-Level DFD FOR Student

This first-level DFD effectively breaks down the student's interactions with the online exam system, showcasing the flow of data between processes and external entities. It reveals the primary functions available to students and provides a clearer understanding of the system's internal processes related to student authentication, exam initiation, profile updates, and result retrieval.

Figure 20 presents a first-level Data Flow Diagram (DFD) focusing on the student's interactions within an online exam system. First-level DFDs decompose the high-level process from the context diagram into its primary sub-processes, revealing more details about the system's internal workings.

This diagram illustrates how a student interacts with the core "Online Exam System", the student first provides their user ID and password to the "Login" process, this information is passed to the "Student" process, which then interacts with the "DataStore" for verification. Upon successful verification, the "Student" process receives confirmation ("Verification Std.") and a signal indicating the student is verified ("Student Verified"), verified student can then engage in three main functionalities:

1. Start Exam: The student initiates an exam ("Start Exam"), and the system provides information about the verified student group ("Verified Student Group") and the selected exam ("Student Exam"). The system also supplies "Question Marks Details" necessary for the exam.
2. Update Profile: The student can update their profile ("Update Prof.") by submitting updated profile information ("Std. Prof.") to the "Student Profile" process.
3. Get Result: The student can retrieve their exam results ("Get Result"). The "Online Exam System" provides the "Result Details."

Figure 20. First-level DFD for student

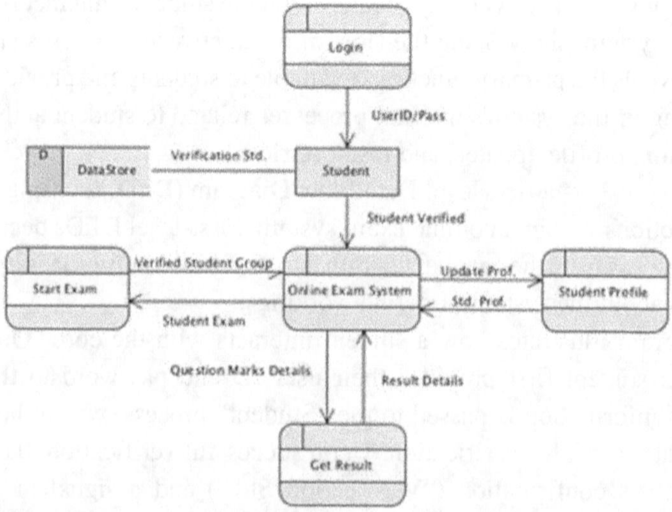

Second-Level DFD

Second-level DFDs delve deeper into specific processes from the first-level diagram, providing a more detailed view of the data flows and transformations within those processes. Figure 21 focuses on the "Online Exam System" process from the first-level DFD, breaking it down into four key sub-processes: "Login," "Online Exam System," "Exam," and "Exam Page View", it represents a second-level DFD for a student within an online exam system, this example includes main key processes:

- Login: The student provides their user ID and password, which are verified against "DataStore2," likely a database containing user credentials. Upon successful verification, the "Login" process outputs "Student Verified."
- Online Exam System: This process manages the overall exam process. It receives "Exam Req." from the student, possibly indicating the student's request to take an exam. The "Online Exam System" then retrieves "Exam Details" from the "Exam" process and sends "Student Details" to the "Student Profile" process. It also receives "Result Details" from the "Result" process and sends them to the "Student" entity.
- Exam: This process handles exam creation and management. It receives "Create Exam" instructions from the "Online Exam System." It also interacts

with the "Exam Page View" process by providing "Exam Page" information and receiving "Exam Page View" data. Additionally, the "Exam" process sets "Question ID" for the exam, which is stored in the "Datasheet" data store.

- Exam Page View: This process likely represents the student's interface for taking the exam. It receives the "Exam Page" from the "Exam" process, displays it to the student, and then sends back the "Exam Page View" information to the "Exam" process. It also interacts with the "QuestionBank" process by sending "Question Req." and receiving "Set Question."

It depicts several data stores, including "DataStore," "Student," "Result," "Questions," and "Datasheet." These data stores represent persistent storage mechanisms where various data related to students, exams, questions, and results are stored and retrieved. The "Result" process interacts with a data store labeled "Result," highlighting the storage of exam results. Similarly, the "QuestionBank" process interacts with the "Questions" data store, indicating the storage and retrieval of questions.

Overall, this second-level DFD provides a more granular view of the student's interactions within the online exam system, outlining the specific data flows and transformations that occur during login, exam management, and exam page interactions.

Figure 21. Second-level DFD for student

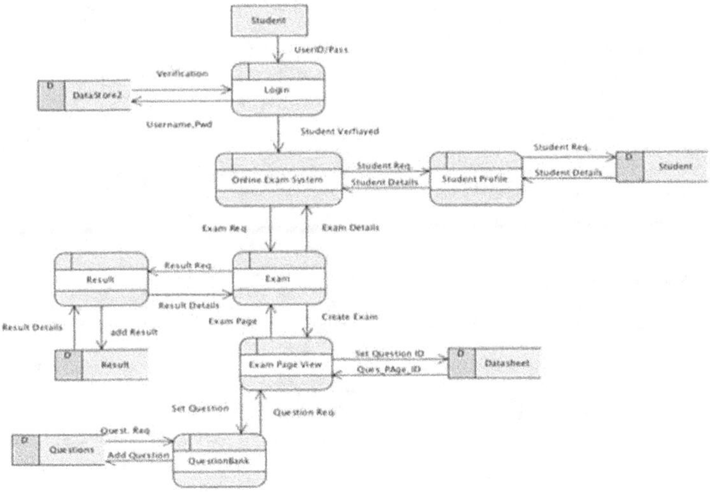

Second-level DFD provides a detailed look at the online exam system from the teacher's perspective. It illustrates the intricate interactions and data flows between the teacher, various system processes, and data stores.

Figure 22 outlines teacher's journey which begins with the "login" process, where they input their User ID and password for verification against the system's "Login" data store. Once authenticated, the teacher can create exams through the "Create Exam" process, providing details and triggering the assignment of the exam to students. The "Create Question" process allows the teacher to add new questions to the system's "QuestionBank," enriching the pool of exam questions. The "Start Exam" process involves the teacher setting exam parameters, including the time, and date, and potentially assigning the exam to specific user groups. After students take the exam, the "Calc Result" process comes into play. It leverages student responses and "UserId with Results" data to generate a comprehensive grade report for the teacher.

Throughout the diagram, data stores like "Login," "Exam," and "QuestionBank" play crucial roles, in storing and providing data required by various processes. The "Teacher," represented as an external entity, actively participates by providing inputs (exam details, question details) and receiving outputs (grade reports). This diagram effectively captures the core functionalities available to teachers within the online exam system, showcasing the flow of information and how teachers interact with the system to create, manage, and assess exams.

Figure 22. Second-level DFD for teacher

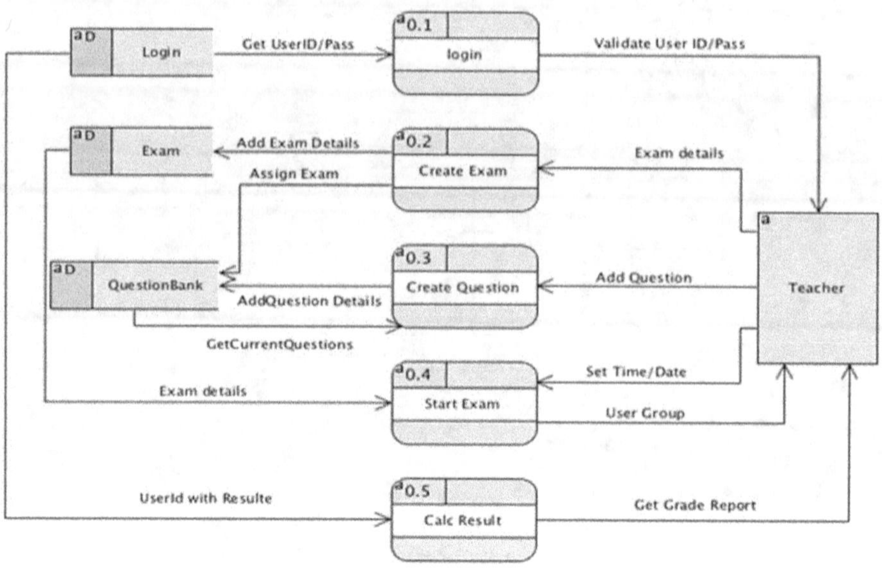

Figure 23 presents a second-level DFD for the administrator, detailing the administrator's role and the data flows involved in managing students, teachers, and exams, the diagram starts with the "Login" process (labeled "0"), where the administrator provides their username and password for verification against the "Admin" data store. Upon successful verification, the "Admin" process receives confirmation ("Verify Admin ID/pass") and the "Admin Verify" signal, allowing them to access the "Exam Online System.". The "Exam Online System" process (labeled "1") serves as the central hub for the administrator's interactions. It receives requests and data from various processes and directs them accordingly.

The administrator has access to three primary functionalities:

1. Manage Student (labeled "2"): The administrator can add new students ("Add Student") or update student details ("Update Details"). This process interacts with the "Student" data store.
2. Manage Teacher (labeled "3"): The administrator can add new teachers ("Add Teacher") or update teacher details. This process interacts with the "Teacher" data store.
3. Manage Exam (labeled "4"): The administrator can add new exams ("Add Exam") and manage existing ones. This process interacts with the "Exam" data store.

The diagram illustrates the flow of data between the administrator, the various processes, and data stores. It highlights the administrator's key role in user management, exam management, and overall system administration. This DFD provides a comprehensive view of the administrator's interactions within the online exam system, emphasizing the system's functionalities from an administrative perspective.

Figure 23. Second-level DFD for administrator

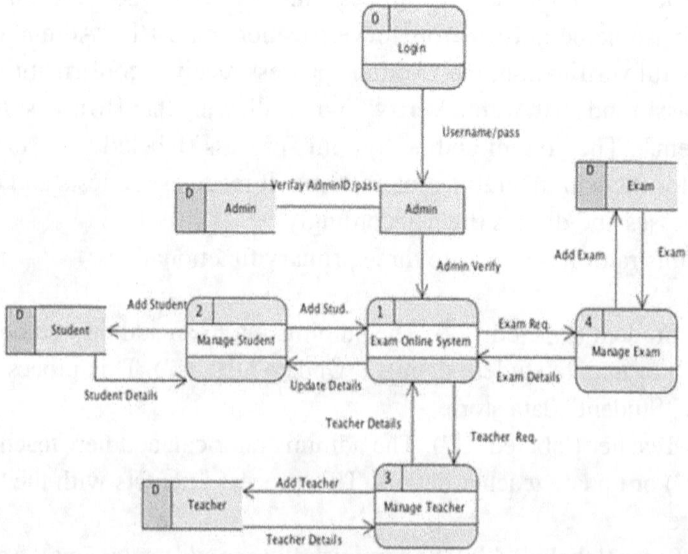

5.2 Entity Relationship Diagram

This ERD effectively outlines the data model for our system, showing the key entities, their attributes, and the relationships that govern their interaction, a graphical representation of the data structure for the system It serves as a blueprint for the database, ensuring efficient data storage, retrieval, and management within the system.

Figure 24 presents an Entity Relationship Diagram (ERD),. The diagram highlights the entities or tables within the database, their attributes, and the relationships between them.

- **Users:** This entity represents users of the system. Attributes include "UserID," "FirstName," "LastName," "Address," "Gender," "UserName," "password," and "GroupID." The "UserID" serves as the primary key, uniquely identifying each user.
- **Exam:** This entity represents exams within the system. Attributes include "ID," "Time," "Questions," "Marks," "ExamResult," "UserID," "GroupID," "Duration," "data," and "finished." The "ID" acts as the primary key for this entity.

- **Result:** This entity stores exam results. Attributes include "ID," "Type," "UserID," "QuestionID," and "Mark." "ID" is the primary key for this entity.
- **Admin:** This entity represents system administrators. Attributes include "ID," "UserName," "StdID," and "UserID." "ID" is the primary key.
- **Question:** This entity represents exam questions. Attributes include "QuestionID," "Question," and "option." "QuestionID" is the primary key.

The relationships between entities are represented by dashed lines with labels indicating the nature of the relationship:

- **Perform:** This relationship connects "Users" and "Exam," indicating that users perform exams.
- **Create:** This relationship connects "Result" to "Admin," suggesting that administrators create result records.
- **Manage:** This relationship connects "Exam" to "Admin," implying that administrators manage exams.
- **Add:** This relationship connects "Admin" to "Question," indicating that administrators add questions to the system.
- **View:** This relationship connects "Users" to "Result," allowing users to view their exam results.

Creating an ERD involves identifying the key entities within the system, their relevant attributes, and the relationships connecting them, below is a step-by-step approach, illustrated using Figure 25, an ERD for an online exam system:

1. determine the key objects or concepts that need to be stored in the database. In Figure 25, the entities are "Users," "Exam," "Result," "Admin," and "Question."
2. Identify the relevant attributes or characteristics. For example, the "Users" entity has attributes like "UserID," "FirstName," "LastName," "Address," and so on.
3. Select a unique attribute or a combination of attributes that uniquely identifies each instance of an entity. In Figure 25, the "UserID" serves as the primary key for the "Users" entity, while "ID" is the primary key for "Exam," "Result," and "Admin."
4. Identify how the entities relate to each other. In Figure 25, the "Users" entity has a "Perform" relationship with the "Exam" entity, indicating that users take exams. The "Result" entity is related to "Admin" through "Create," suggesting that administrators create result records. Other relationships include "Manage," "Add," and "View."

5. Define the number of instances of one entity that can be associated with instances of another entity. For example, a one-to-many relationship (1:N) between "Exam" and "Result" indicates that one exam can have multiple results.

Figure 24. Entity relationship diagram

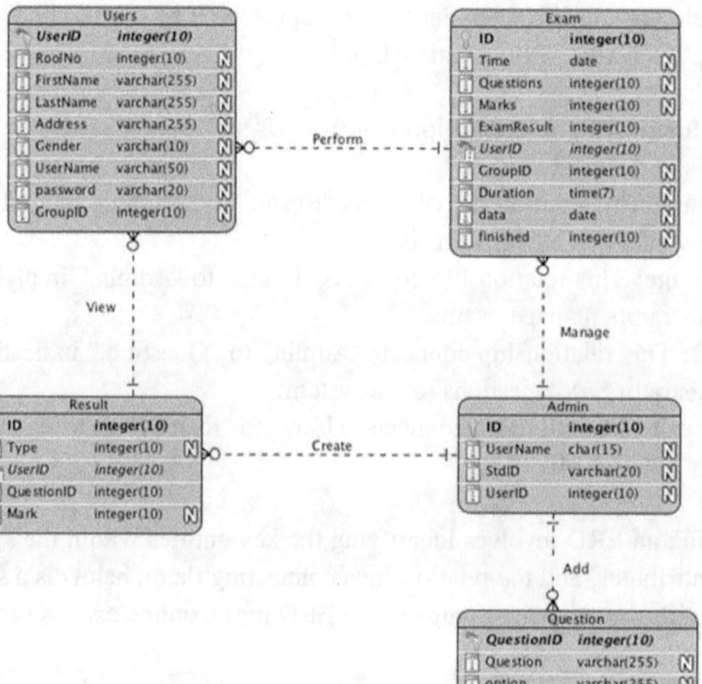

Figure 25 effectively captures the data structure of the online exam system, showcasing the key entities, their attributes, and the relationships that govern their interaction. This ERD serves as a blueprint for the database, enabling efficient data management and ensuring data integrity and consistency.

CONCLUSION

This chapter has provided a comprehensive overview of the essential UML diagrams and concepts for designing an online exam system. We explored detailed design class diagrams, dynamic models (sequence diagrams, activity diagrams, and state diagrams), software architecture diagrams (package diagrams and component diagrams), and Entity Relationship Diagrams. By applying these techniques, we can effectively translate abstract requirements into a well-structured, robust, and maintainable software design. The diagrams presented in this chapter serve as blueprints for developers, offering a clear understanding of the system's structure, behavior, and data management. This detailed design phase lays the groundwork for the successful implementation and deployment of the AOES system.

Moving forward, the insights gained from these diagrams will be crucial for guiding the development team. The clear understanding of the system's architecture, data structure, and dynamic behavior will ensure that the implementation aligns with the design specifications and meets the functional and non-functional requirements. The visual nature of these diagrams facilitates communication between stakeholders, ensuring a shared understanding of the system's design and functionalities. The insights gained during the design phase will streamline the development process, promoting efficiency, reducing errors, and ultimately contributing to the successful realization of the AOES system.

Finally, we bridge the gap between software design and database implementation by exploring Entity Relationship Diagrams (ERDs). These diagrams map the data structures defined in class diagrams to the corresponding database schema, ensuring a seamless connection between the software design and the underlying data storage. The ERD demonstrates how entities like users, exams, results, administrators, and questions are represented in the database, along with their attributes and relationships. The diagram highlights the importance of primary keys, cardinality, and different types of relationships in designing an effective database schema.

REFERENCES

Davis, W. S. (2019). Data flow diagrams. *The Information System Consultant's Handbook*, 175–188. 10.1201/9781420049107-24

Ľachová, K., & Trebuňa, P. (2019). Modeling of electronic kanban system by using entity relationship diagrams. *Acta Logistica*, 6(3), 63–66. Advance online publication. 10.22306/al.v6i3.115

Zhang, H., Liu, W., Xiong, H., & Dong, X. (2018). Analyzing data flow diagrams by combination of formal methods and visualization techniques. *Journal of Visual Languages and Computing*, 48, 41–51. Advance online publication. 10.1016/j.jvlc.2018.08.001

Section 3
Implementation and Testing

Final implementation chapter is the last stage of your software development, this state which parts of the requirements are met and which are not. Furthermore, the implementation section include the user and installation guides for your system too. This section explain the server technologies and management services associated with the hosting and management of secure websites, development frameworks, website publishing, content management, search engine optimization, tools and software used to develop websites, utilize website technologies, tools and techniques with good design principles to deploy a multipage website and create and use a Test Plan to review the performance and design of a multipage website.

Chapter 5
Implementing AOES Website Using PHP

ABSTRACT

The implementation phase is the last stage of your software development, in which you will have the chance to describe any difficulties you have faced during the implementation. You must also state which parts of the requirements are met and which are not. Furthermore, the implementation part must include the user and installation guides for your system, too. Software implementation is the third stage in which you are supposed to implement the software system you have designed in the previous stage. You can use any object-oriented programming languages to implement your system, e.g., PHP and C#. In this chapter, the authors applied software engineering techniques for AOES example to implement a real case to develop a personal website with all user requirements and show in detail how they transfered all system analysis to the whole system as a real practical case.

INTRODUCTION

This chapter takes us to implementation stage of software development, we will transform the abstract concepts and designs from previous chapters into a tangible, working ABC online examination system (AOES), we will use PHP, a popular server-side scripting language, and MySQL, a relational database management system, to build a functional web application that meets the requirements defined in the analysis phase. We will provide a comprehensive walkthrough of the implementation process, covering each user role (administrator, teacher, and student) and their corresponding functionalities. The chapter will include code snippets,

DOI: 10.4018/978-1-6684-4844-1.ch005

screenshots of the user interface, and explanations of key features. We will address the challenges faced during implementation and discuss how we resolved them.

The goal of this chapter is to provide readers with a practical example of how to implement a real-world web application using software engineering principles and UML diagrams as guides. By the end of this chapter, readers will have a solid understanding of the implementation phase and the ability to apply these techniques to develop their own online exam systems or similar web applications.

AOES is web application provide question to the student randomly should include three type of users (student, teacher and administrator), the main function for the system to create the exam, display to students to fill and submit and get result, exam timer should be show once the student start Exam attempts, the system should satisfy the three sub modules, admin will have use all the functions that available to students and teacher, teacher can use student function after admin allow or disallow in user roll table, student can access the system and start his exam. This chapter is proposing all features and procedures to implement the system for Abc University that will apply for English teacher. System implementation phase is the 3rd phase after system analysis and design phases, so when we start in this level, we use all details that describe in analysis and design documentation that used in previous chapter.

In this chapter we will start to describe the implementation phase (Moreb et al., 2020; *SEMLHI*, n.d.)and describing difficulties that we have faced with during the implementation, it is very important that not all the requirement we implement in the system related to the state on the system in implementation and some data that need from deployment environment in the Abc University, figure 1 shown the implementation plane for our practical example as illustrate in Figure 1.

Figure 1. Implementation plane

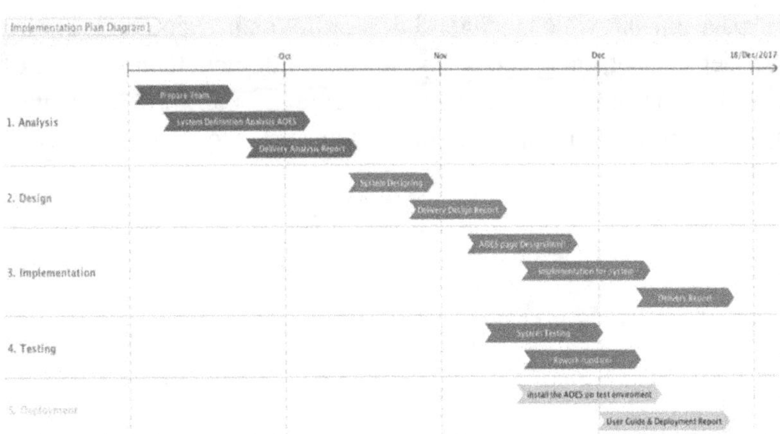

In this chapter we described the sequence that the user should be followed to use the system based on previous chapter for system analysis, as you recall there are three main users in the system (administrator, teacher, and student), and to use this system, it is very import to know about the different users and their roles which are given below.

1. Administrator: Super user, adds teacher and manages system, do all admin task and access all system feature.
2. Teacher: Another user name as teacher member, lecturer or examiner who create exam and questions with answers and can access grad book.
3. Student: User mostly a student who will appear for the examination.
4. System: the system do automatic task that assign by user and should execute depend on time or events.

1. AOES MAIN PAGE

After the user open website using internet browser (Chrome, Firefox, Microsoft Exp.) the system main page display to user, the main page include information about the system and announcement, university news all news related to system user can found in this page, then depending on the user role user will select to login to the system or if the user is new will fill the registration form then the system will automatically cheek if the student have done all the administration steps and pay the fees, if yes the student will login and can start the exam.

The AOES Main Page include three parts as following:

1. main page top part which include slideshow for main events and activity related to the department, this page design include slideshow then three box for user login (student and teacher) as shown in Figure 2, for any problem that user need help they can click on help disk to submit this ticket to resolve by administrator.

Figure 2. Main page top part

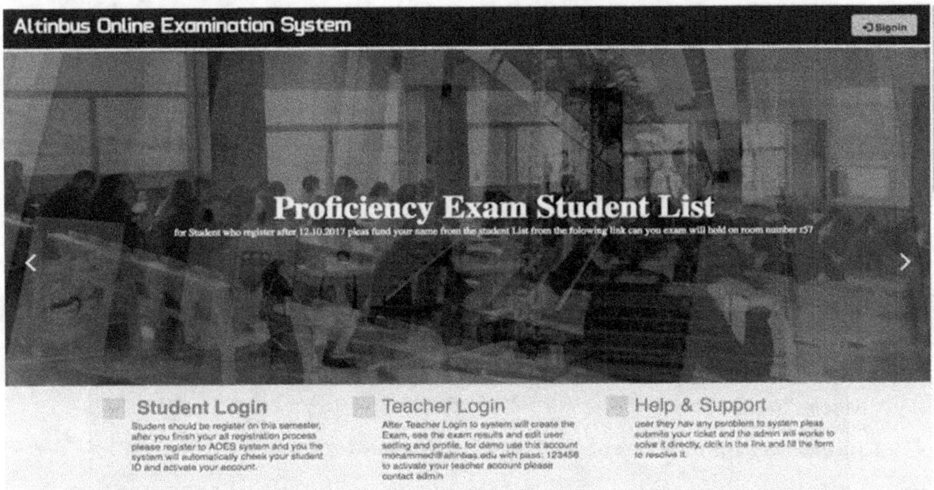

2. main page part includes the registration form for new user as shown in Figure 3, it include the essential steps for new user that the user should follow before to start exam.

Figure 3. Main page part

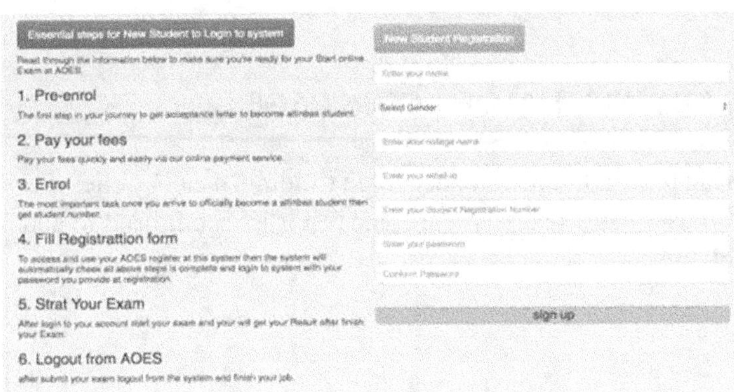

3. news and announcement part which display all upcoming event and exam session, quick links and main news for university related to news student, and at the footer of the page include main information about the system with admin link to login to admin panel as shown in Figure 4.

Figure 4. News and announcement part

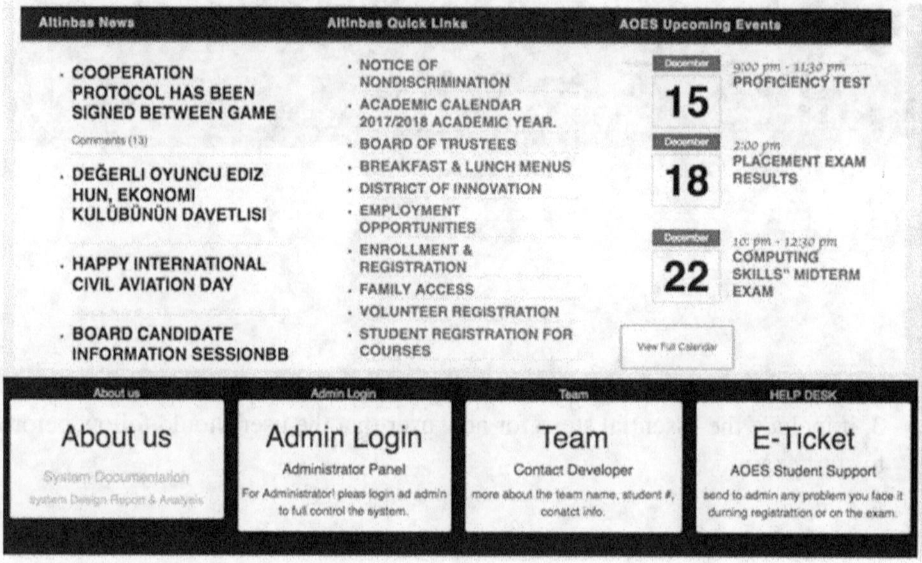

2. AOES ADMINISTRATOR LOGIN PAGE

From main page, administrator can login to admin panel by click on admin login link then fill user name and password, for test it is recommend to add account as shown in Figure 5.

Figure 5. Administrator login page

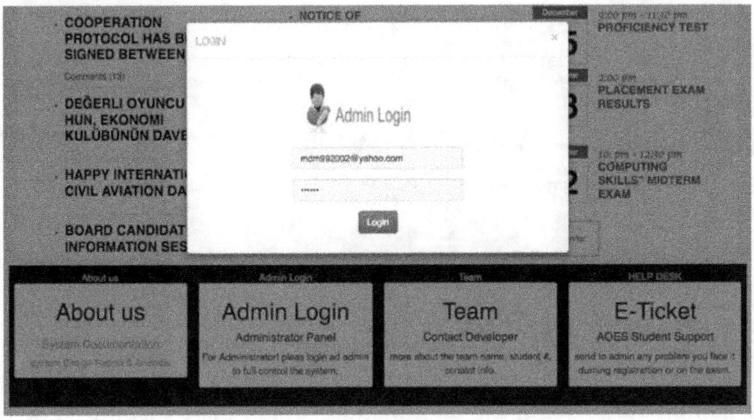

if admin add not correct credential data error massage will display to user as shown in Figure 6.

Figure 6. Error message for non-admin account

Details for main page implementation code used:

- User login form.
- query string passed to admin page.
- Validate user name and password and if correct display admin page

Administrator Login Form Code

```
<form role="form" method="post" action="admin.php?q=index.php">
<div class="form-group">

  <img src="image/adminlogin.JPG" width="200" height="120" alt=""/>
  <input type="text" name="uname" maxlength="20"  placeholder="Admin user id" class="form-control"/>
  </div>
<div class="form-group">
  <input type="password" name="password" maxlength="15" placeholder="Password" class="form-control"/>
  </div>
<div class="form-group" align="center">
  <input type="submit" name="login" value="Login" class="btn btn-primary" />
  </div>
  </form>
```

3. AOES ADMINISTRATOR MAIN PAGE

After administrator login the administrator main page display as below, main function and feature admin can use in this page as following:

- Manage Exam
  - Add new Exam
  - Remove Exam
  - View published Exam.
  - Start Exam.
  - Restart Exam
- Manage Users
  - View users profile
  - View admin Profile
  - Delete user account
  - logout
- Grade book
- Help Disk
  - View user ticket massage
  - Replay to user massage
  - Delete ticket.
  - Help
- Calendar

Figure 7. Administrator main page

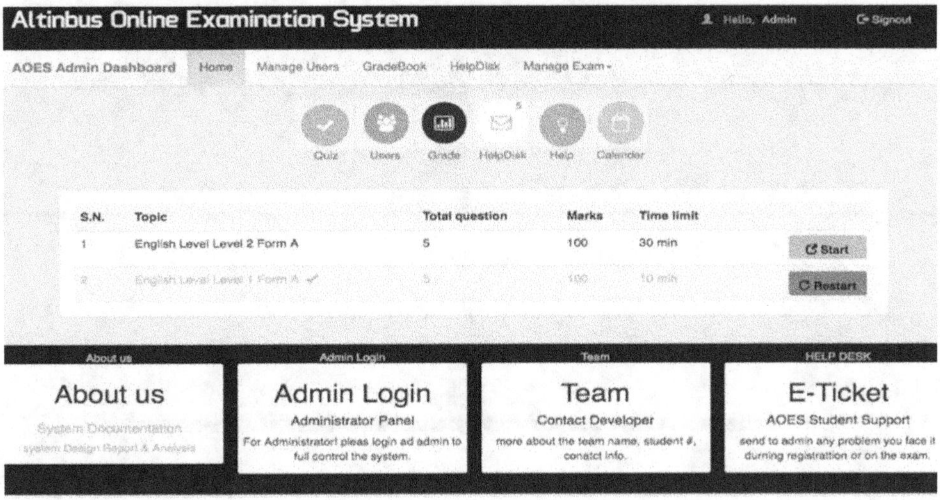

The home admin page displays the exam that created and published to students, admin can start the exam and view the questions.

3.1 Add New Exam Page

This screen allow admin to create new exam include the following data:

- Exam Title
- The total question that the Exam include.
- Marks for correct questions.
- Marks for wrong answer.
- Exam time in minutes.
- Exam nots or tags for search option.
- Exam descriptions and details.

After the data add user click on submit button to go to next screen to add the questions for the exam, we add in this system one question type (multiple answer).

Figure 8. Add new exam page

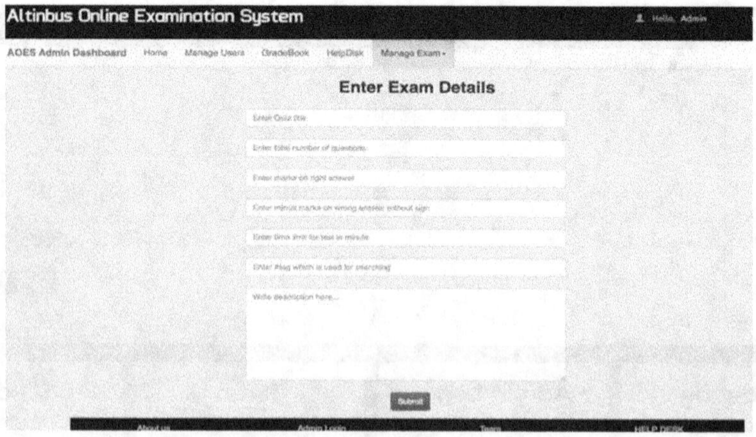

php class used to create the exam, after the user click on submit button on add new exam the below code save input value and add to MySQL database (West & Prettyman, 2018).

```
//add quiz
if(isset($_SESSION['key'])){
if(@$_GET['q']== 'addquiz' && $_SESSION['key']=='mmoreb2017') {
$name = $_POST['name'];
$name= ucwords(strtolower($name));
$total = $_POST['total'];
$sahi = $_POST['right'];
$wrong = $_POST['wrong'];
$time = $_POST['time'];
$tag = $_POST['tag'];
$desc = $_POST['desc'];
$id=uniqid();
$q3=mysqli_query($con,"INSERT INTO quiz VALUES  ('$id','$name' , '$sahi' , '$wrong','$total','$time' ,'$desc','$tag',
NOW())");

header("location:dash.php?q=4&step=2&eid=$id&n=$total");
}
}
```

3.2 Add Question Details

Add question page as shown in Figure 9, allow admin to add new questions include the following data for each question:

- Question text.
- Option A.
- Option B.
- Option C.
- Option D.
- Correct answer.

Figure 9. Add question details

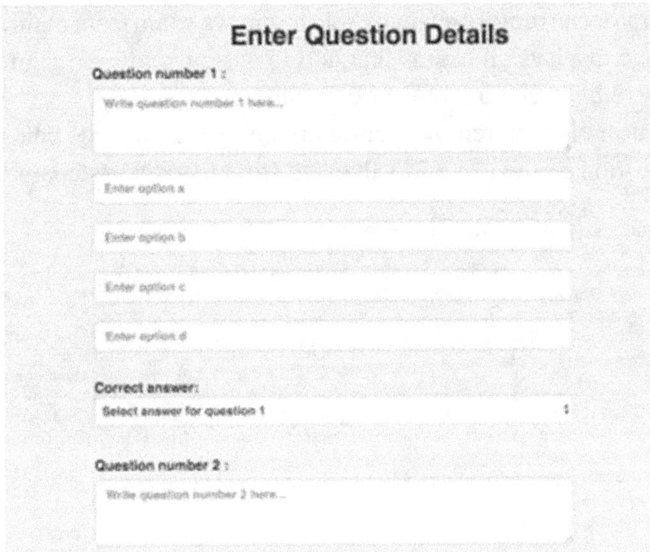

After the user click on submit button on add new questions php class call the below code which save input value and add to MySQL database.

```
if(isset($_SESSION['key'])){
if(@$_GET['q']== 'addqns' && $_SESSION['key']=='mmoreb2017') {
$n=@$_GET['n'];
$eid=@$_GET['eid'];
$ch=@$_GET['ch'];

for($i=1;$i<=$n;$i++)
 {
 $qid=uniqid();
 $qns=$_POST['qns'.$i];
$q3=mysqli_query($con,"INSERT INTO questions VALUES  ('$eid','$qid','$qns' , '$ch' , '$i')");
  $oaid=uniqid();
  $obid=uniqid();
$ocid=uniqid();
$odid=uniqid();
$a=$_POST[$i.'1'];
$b=$_POST[$i.'2'];
$c=$_POST[$i.'3'];
$d=$_POST[$i.'4'];
$qa=mysqli_query($con,"INSERT INTO options VALUES  ('$qid','$a','$oaid')") or die('Error61');
$qb=mysqli_query($con,"INSERT INTO options VALUES  ('$qid','$b','$obid')") or die('Error62');
$qc=mysqli_query($con,"INSERT INTO options VALUES  ('$qid','$c','$ocid')") or die('Error63');
$qd=mysqli_query($con,"INSERT INTO options VALUES  ('$qid','$d','$odid')") or die('Error64');
$e=$_POST['ans'.$i];
switch($e)
{
case 'a':
$ansid=$oaid;
break;
case 'b':
$ansid=$obid;
break;
case 'c':
$ansid=$ocid;
break;
case 'd':
$ansid=$odid;
break;
default:
$ansid=$oaid;}
$qans=mysqli_query($con,"INSERT INTO answer VALUES  ('$qid','$ansid')");}
header("location:dash.php?q=0");}}
```

3.3 Remove Exam

Administrator can remove exam by select remove exam from main administrator page, this page displays all exams that add to system and user can remove which exam to remove by click on remove exam button.

After the user click on remove button on remove Exam page below code delete exam and question by the exam id value and from MySQL database.

```
//remove quiz
if(isset($_SESSION['key'])){
if(@$_GET['q']== 'rmquiz' && $_SESSION['key']=='mmoreb2017') {
$eid=@$_GET['eid'];
$result = mysqli_query($con,"SELECT * FROM questions WHERE eid='$eid' ") or die('Error');
while($row = mysqli_fetch_array($result)) {
     $qid = $row['qid'];
$r1 = mysqli_query($con,"DELETE FROM options WHERE qid='$qid'") or die('Error');
$r2 = mysqli_query($con,"DELETE FROM answer WHERE qid='$qid' ") or die('Error');
}
$r3 = mysqli_query($con,"DELETE FROM questions WHERE eid='$eid' ") or die('Error');
$r4 = mysqli_query($con,"DELETE FROM quiz WHERE eid='$eid' ") or die('Error');
$r4 = mysqli_query($con,"DELETE FROM history WHERE eid='$eid' ") or die('Error');

header("location:dash.php?q=5");
}
}
```

3.4 Manage User Page

This screen allow admin to Manage users include the following data:

- Student number
- Full Name
- Gender
- College
- Email
- Mobile
- Action (delete, update, Activate).

Administrator can view all student registered to system and can manage user by delete or disable account by click the delete icon on the action column as shown in Figure 10.

Figure 10. Mange users

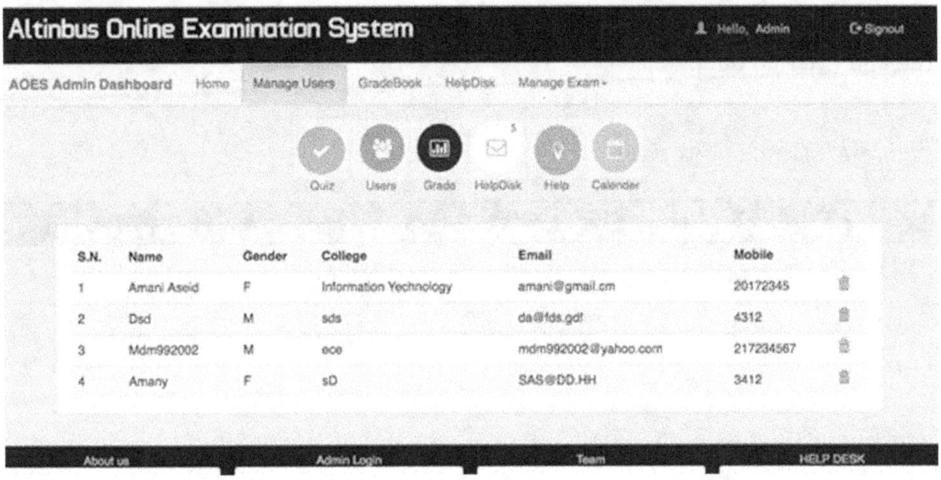

After the user click on delete button on user page below code delete user and profile data by the user id value from MySQL database.

```
//delete user
if(isset($_SESSION['key'])){
if(@$_GET['demail'] && $_SESSION['key']=='mmoreb2017') {
$demail=@$_GET['demail'];
$r1 = mysqli_query($con,"DELETE FROM rank WHERE email='$demail' ") or die('Error');
$r2 = mysqli_query($con,"DELETE FROM history WHERE email='$demail' ") or die('Error');
$result = mysqli_query($con,"DELETE FROM user WHERE email='$demail' ") or die('Error');
header("location:dash.php?q=1");
}
}
```

3.5 Grade Book Page

This centralized view allows the administrator to easily assess and compare student performance. Additionally, the administrator (or teacher, with appropriate permissions) can export this gradebook data. The current export functionality involves copying the table content and pasting it into an external spreadsheet application like Excel, this enables further data manipulation, filtering, and analysis as needed. This screen allow admin to view users grade the following data:

- Rank
- Full Name
- Gender
- College
- Score

Administrator can view all student grade on one table for selected exam as shown in Figure 11, as report which include student rank with other student, to export the gradebook admin or teacher can copy the content of table and past data on excel file and filter all data.

Figure 11. Grade book page

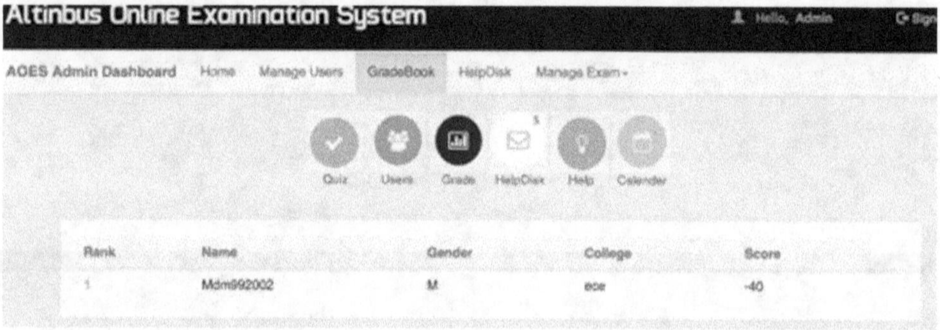

3.6 Help Disk Page

Help disk page include simple ticket system allow user to submit ticket or massage and then admin read the massage and submit replay using contact information that the user include when submit massage. Help disk include the following data:

- Massage number
- Subject
- Email
- Date
- Time
- By
- Action(read, delete)

Administrator can view all user massage(ticket) on one table as shown in Figure 12, main user info show on the table and admin can read the massage by click on read icon or delete by click on delete massage icon.

Figure 12. Help desk page

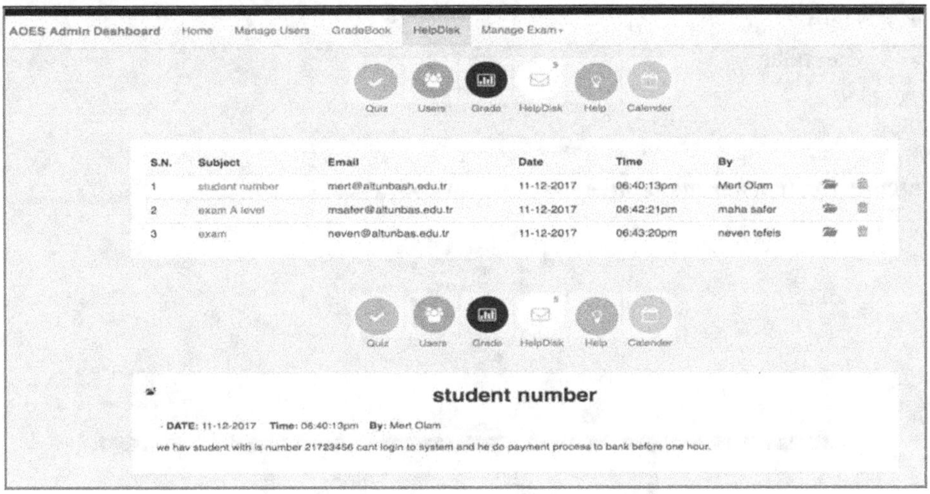

php code call after admin read massage can delete it by massage id.

```
if(isset($_SESSION['key'])){
if(@$_GET['fdid'] && $_SESSION['key']=='mmoreb2017') {
$id=@$_GET['fdid'];
$result = mysqli_query($con,"DELETE FROM feedback WHERE id='$id' ") or die('Error');
header("location:dash.php?q=3");
}
}
```

4. AOES TEACHER MAIN PAGE

The teacher includes most of the privilege that administrator can do without manage user, add teacher, after teacher login the teacher main page display as below, code that use for teacher page we use the same php class that use in administrator page, main function and feature teacher can use in this page as following:

- Manage Exam
    - Add new Exam
    - Remove Exam
    - View published Exam.
    - Start Exam.
    - Restart Exam
- Users
    - View users profile
    - View teacher Profile

- o logout
- Grade book
- Help
- Calendar

Figure 13. Teacher main page

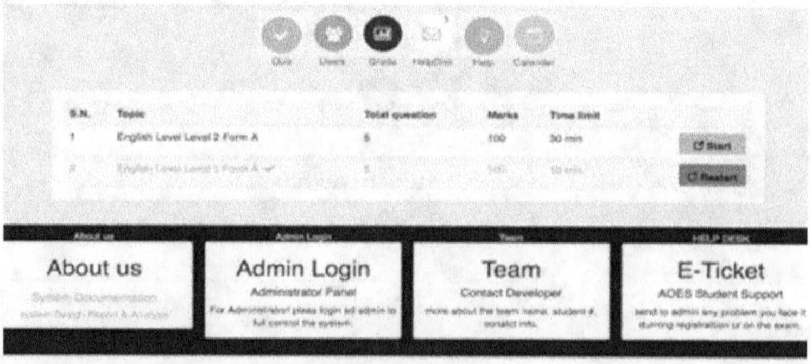

Teacher home page as shown in Figure 13 display the exam that created and published to students, teacher can start the exam and view the questions.

4.1 Add New Question Page

Add new question page screen allow teacher to create new exam include the following data:

- Exam Title
- The total question that the Exam include.
- Marks for correct questions.
- Marks for wrong answer.
- Exam time in minutes.
- Exam nots or tags for search option.
- Exam descriptions and details.

After the data add user click on submit button to go to next screen to add the questions for the exam, we add in this system one question type (multiple answer) as shown in Figure 14.

Figure 14. Add new exam page

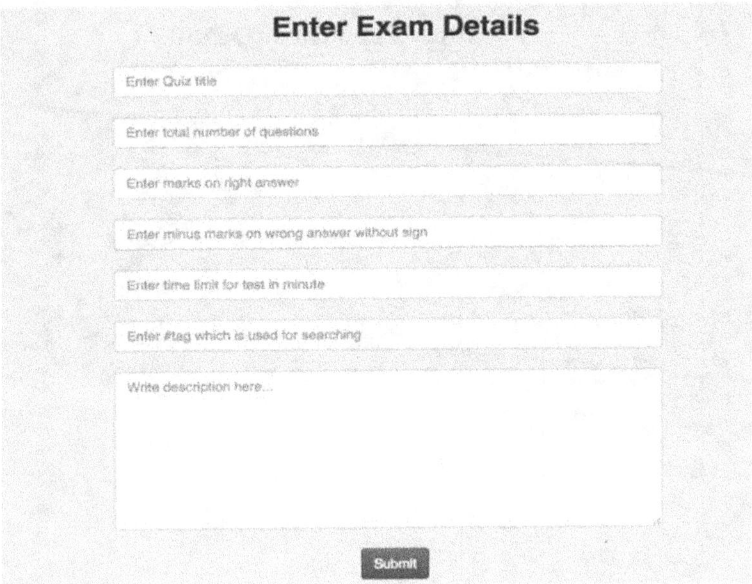

4.2 Add Question Details.

Add question page as shown in Figure 15, allow teacher to add new questions include the following data for each question:

- Question text.
- Option A.
- Option B.
- Option C.
- Option D.
- Correct answer.

Figure 15. Add question details

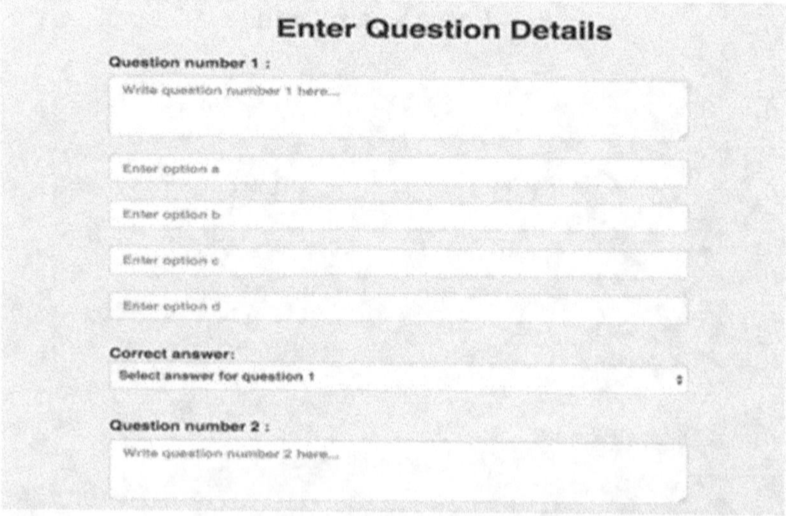

4.3 Remove Exam

Teacher can remove exam by select remove exam from main display table, this page displays all exams that add to system and user can remove exam by click on remove exam button as shown in Figure 16.

Figure 16. Remove exam

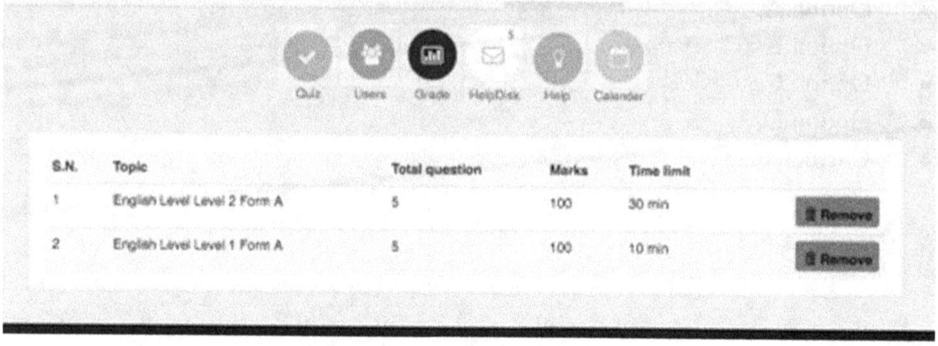

4.4 Grade Book page

This screen allow teacher to view users grade, the gradebook include the following data:

- Rank
- Full Name
- Gender
- College
- Score

Teacher can view all student grade on one table for selected exam as shown in Figure 17, the report include student rank with other student, to export the gradebook teacher can copy the content of table and past data on excel file and filter all data as shown in Figure 17.

Figure 17. Grade book page

Rank	Name	Gender	College	Score
1	Mdm992002	M	ece	-40

5. AOES STUDENT MAIN PAGE

Main steps that system allow the student to start exam by student should follow through the information below to make sure your ready login as follow.

1. Pre-Enrollment: Apply and receive an acceptance letter to become register student, it is first step in your journey to get acceptance letter to register as student to system.
2. Fee Payment: Pay your fees conveniently through AOES online payment service.
3. Enrollment: Officially enroll as an AOES student and obtain your student number.
4. AOES Registration: Register for an AOES account. The system will automatically verify your enrollment status and grant you access. Use the password you provide during registration to log in.

5. Take Your Exam:
    a. Log in to your AOES account.
    b. Navigate to the Exam section.
    c. Select the exam you are scheduled to take.
    d. Review the exam instructions carefully.
    e. Click Start Exam to begin.
    f. Answer all questions within the allotted time.
    g. Submit your exam when finished. Your results will be available upon completion.
6. Logout: Remember to log out of your AOES account after finishing your exam.

Figure 18 Shown how we implement all system requirements and use case to implement real page, the home student page display the exam that available and published to students, student can start the exam and view the questions.

Figure 18. Student main page

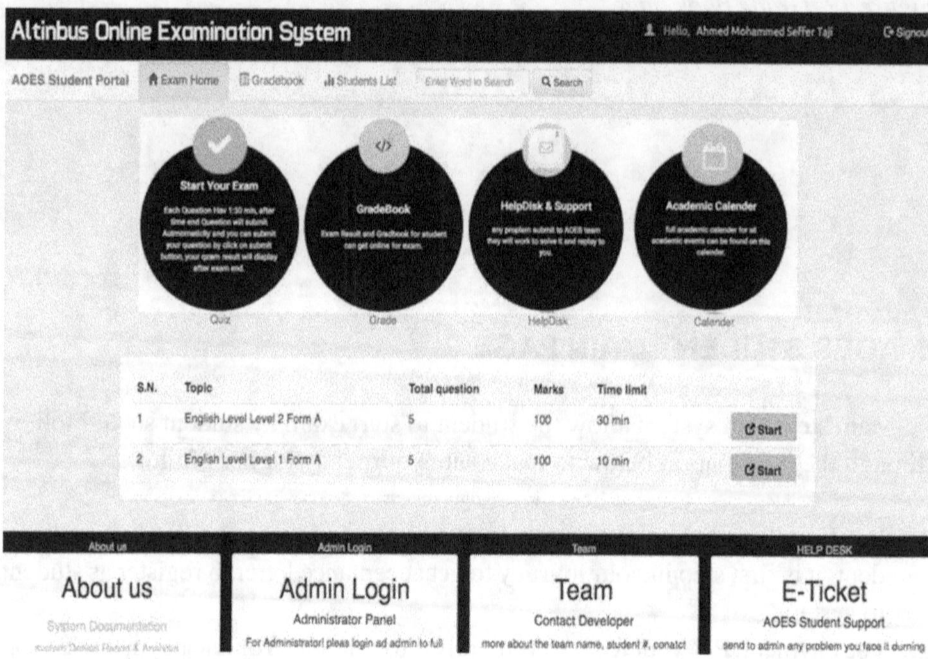

5.1 Student Registration Page

This screen allow student to register to AOES from the main page, new student registration form include the following data:

- Student full name
- Gender.
- College name.
- Email address.
- Student Registration number.
- password.

After the data add user click on sign up button to go validate data add, we assume that our AOES connected with registration database to select in user he finishes the first 3 step for registration, after user data add to database system will automatically login student and display published exam.

Figure 19. Student registration page

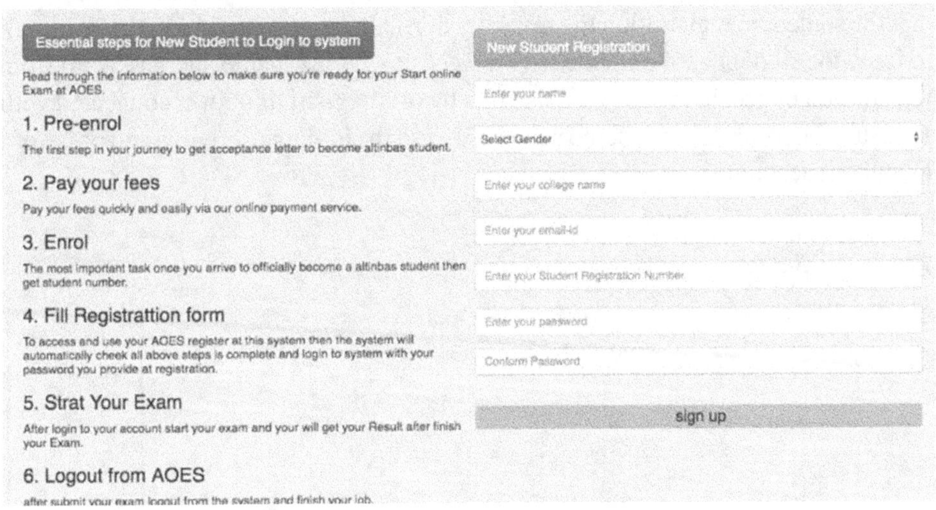

php class to create exam which used after the user click on sign up button on registration form which save input value and add to MySQL database.

```php
<?php
include_once 'dbConnection.php';
ob_start();
$name = $_POST['name'];
$name= ucwords(strtolower($name));
$gender = $_POST['gender'];
$email = $_POST['email'];
$college = $_POST['college'];
$mob = $_POST['mob'];
$password = $_POST['password'];
$name = stripslashes($name);
$name = addslashes($name);
$name = ucwords(strtolower($name));
$gender = stripslashes($gender);
$gender = addslashes($gender);
$email = stripslashes($email);
$email = addslashes($email);
$college = stripslashes($college);
$college = addslashes($college);
$mob = stripslashes($mob);
$mob = addslashes($mob);

$password = stripslashes($password);
$password = addslashes($password);
$password = md5($password);

$q3=mysqli_query($con,"INSERT INTO user VALUES  ('$name' , '$gender' , '$college','$email' ,'$mob', '$password')");
if($q3)
{
session_start();
$_SESSION["email"] = $email;
$_SESSION["name"] = $name;

header("location:account.php?q=1");
}
else
{
header("location:index.php?q7=Email Already Registered!!!");
}
ob_end_flush();
?>
```

5.2 Student Login Page

This screen allow student to login to AOES from the main page username should be email address and password that provide in registration form, we should notes that the student in registration to university normally process need one to two month to issue the student card and one to two weeks to get username and password from registration office, in AOES we solve this by user registration form connected with Registration office DB, for demo account we use the fooling username and password as student account:

Figure 20. Student login page

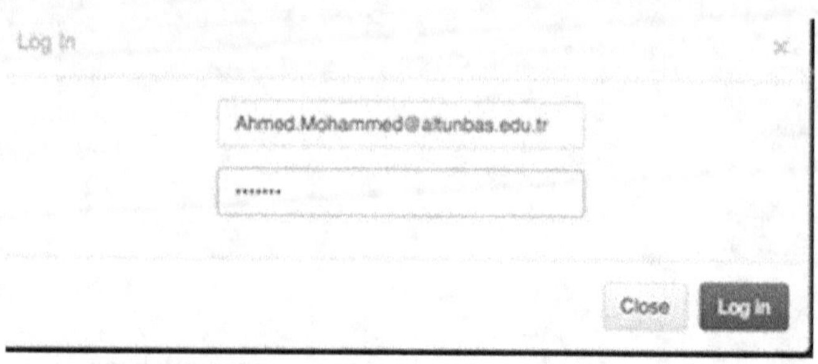

If student add incorrect data, a massage will show in box massage.

5.3 Student Exam Main Page

After the student login to system student can select from main page Exam to start it as shown in Figure 21, exam table display the following data for published exam:

- Exam number
- Topic
- Total question
- Marks
- Time limit
- Action (start exam)

Figure 21. Student exam main page

5.4 Start Exam Page

Student can Start exam after login to system by select Start Exam from action column, after student click on start the question display each page include one question and the time for each question maximum 1:30 minutes, if user Exceed the time the question will submit automatically and next question display to student until finish the exam the result will show to user, later on the user can login to system and see exam result from grade book menu.

Figure 22. Exam question page

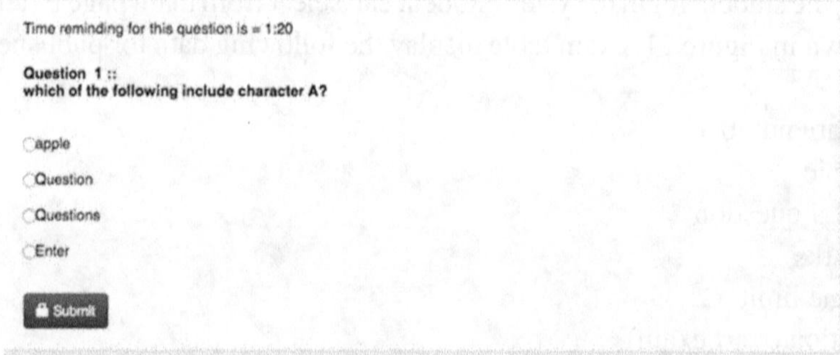

After all question summited or time finish for exam, student will get result as next report.

Figure 23. Result report

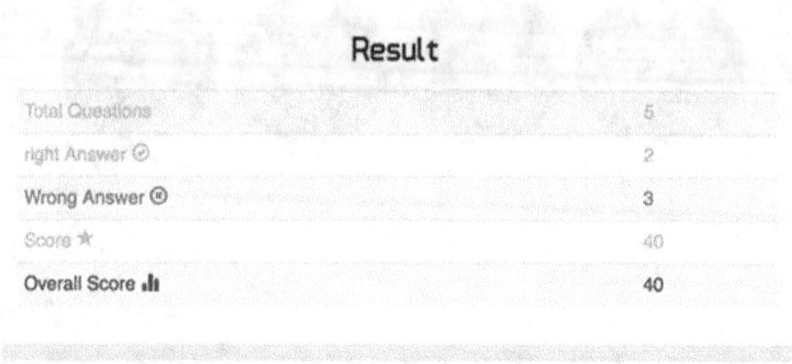

After the user click on start exam button on student exam page, the below code display question on page and save result to MySQL database.

```
if(@$_GET['q']== 'quiz' && @$_GET['step']== 2) {
$eid=@$_GET['eid'];
$sn=@$_GET['n'];
$total=@$_GET['t'];
$ans=$_POST['ans'];
$qid=@$_GET['qid'];
$q=mysqli_query($con,"SELECT * FROM answer WHERE qid='$qid' " );
while($row=mysqli_fetch_array($q) )
{
$ansid=$row['ansid'];
}
if($ans == $ansid)
{
$q=mysqli_query($con,"SELECT * FROM quiz WHERE eid='$eid' " );
while($row=mysqli_fetch_array($q) )
{
$sahi=$row['sahi'];
}
if($sn == 1)
{
$q=mysqli_query($con,"INSERT INTO history VALUES('$email','$eid' ,'0','0','0','0',NOW())")or die('Error');
}
$q=mysqli_query($con,"SELECT * FROM history WHERE eid='$eid' AND email='$email' ")or die('Error115');

while($row=mysqli_fetch_array($q) )
{
$s=$row['score'];
$r=$row['sahi'];
}
$r++;
$s=$s+$sahi;
$q=mysqli_query($con,"UPDATE `history` SET `score`=$s,`level`=$sn,`sahi`=$r, date= NOW()  WHERE  email = '$email' AND eid =
'$eid'")or die('Error124');

}
else
{
$q=mysqli_query($con,"SELECT * FROM quiz WHERE eid='$eid' " )or die('Error129');

while($row=mysqli_fetch_array($q) )
{
$wrong=$row['wrong'];
}
```

Timer script used on the exam question page.

```
<script>
var seconds = 90;
      function secondPassed() {
          var minutes = Math.round((seconds - 30)/60),
              remainingSeconds = seconds % 60;

          if (remainingSeconds < 10) {
              remainingSeconds = "0" + remainingSeconds;
          }

          document.getElementById('countdown').innerHTML = minutes + ":" + remainingSeconds;
          if (seconds == 0) {
              clearInterval(countdownTimer);
             //form1 is your form name
            document.formquize.submit();
          } else {
              seconds--;
          }
      }
      var countdownTimer = setInterval('secondPassed()', 1000);
    </script>
```

5.5 Gradebook Page

Student can get exam result by click on gradebook icon from student main page as shown in Figure 21, gradebook table include the following data:

- Exam Number
- Exam(Quiz) Title
- Question Solved
- Right answer
- Wrong answer
- Total marks (Score)

This page displays all student who participate in exam, this page display main data for student whom apply to exam, participant student table include the following data:

- Rank for student

- Full Name
- Gender
- College
- Score

Figure 24. Student list page

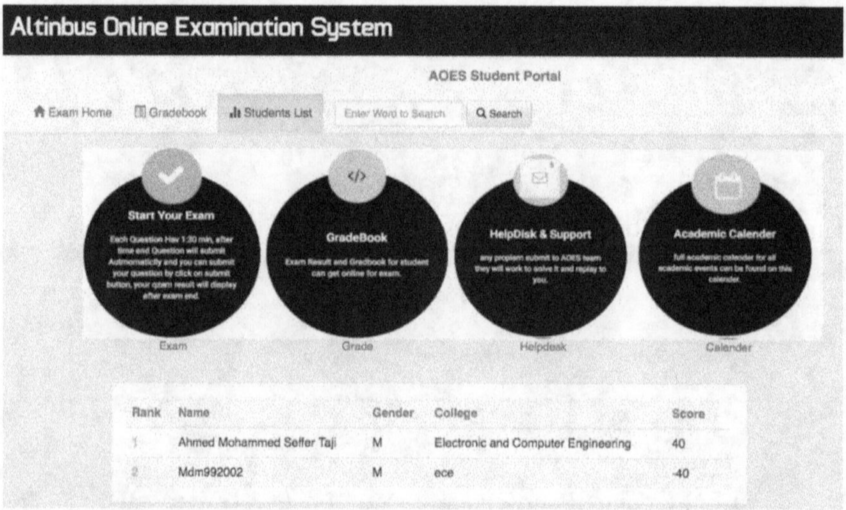

5.7 Helpdesk Page

From student main page as shown in Figure 25, for help and support student can click on Helpdesk icon or menu then fill description and contact information then submit to admin to solve it, student Helpdesk form include the following data:

- Name.
- Subject.
- E-Mail address:
- Description for your problem.

Figure 25. Helpdesk page

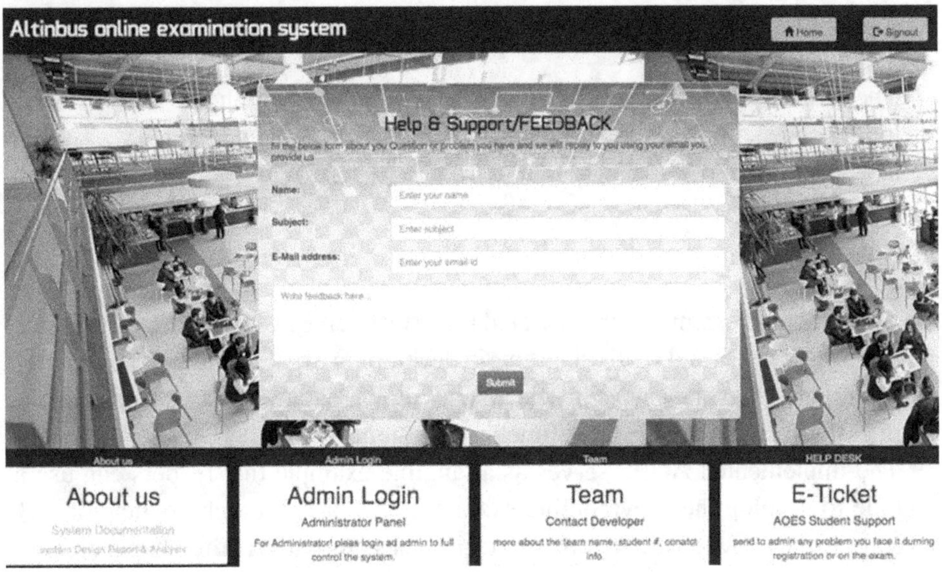

After the user clicks on submit button on Helpdesk page php class call listed below, this code save result to MySQL database.

```
<?php
include_once 'dbConnection.php';
$ref=@$_GET['q'];
$name = $_POST['name'];
$email = $_POST['email'];
$subject = $_POST['subject'];
$id=uniqid();
$date=date("Y-m-d");
$time=date("h:i:sa");
$feedback = $_POST['feedback'];
$q=mysqli_query($con,"INSERT INTO feedback VALUES  ('$id' , '$name', '$email' , '$subject', '$feedback' , '$date' , '$time')")or die
("Error");
header("location:$ref?q=Thank you for your valuable feedback");
?>
```

RESULT

This chapter detailed the implementation of an online exam system (AOES) using PHP and MySQL. We successfully transformed the analysis and design from previous chapters into a functional web application, realizing the core functionalities for each user role: administrator, teacher, and student. The implemented system includes:

- User authentication and registration for all roles.
- Exam creation and management capabilities for teachers and administrators.
- Secure question management and storage in a question bank.
- Exam delivery with a timer to enforce time limits.

- Automated result calculation and presentation.
- Gradebook management for teachers and administrators.
- A help desk system for user support.

The chapter provided a step-by-step walkthrough of the implementation process, including code snippets, screenshots, and explanations of key features.

CONCLUSION

This practical example demonstrated the effectiveness of applying software engineering principles and UML diagrams in developing a real-world web application, we highlighted how the analysis, design, and implementation phases work together seamlessly to create a robust and functional system.

The implemented AOES serves as a tangible example that readers can use as a guide to develop their own online exam systems or other web applications. By leveraging the knowledge and practical techniques presented in this chapter, readers can confidently tackle real-world software development challenges.

REFERENCES

Moreb, M., Mohammed, T. A., & Bayat, O. (2020). A Novel Software Engineering Approach Toward Using Machine Learning for Improving the Efficiency of Health Systems. *IEEE Access : Practical Innovations, Open Solutions*, 8, 23169–23178. 10.1109/ACCESS.2020.2970178

West, A. W., & Prettyman, S. (2018). *Practical PHP 7, MySQL 8, and MariaDB Website Databases*. 10.1007/978-1-4842-3843-1

Chapter 6
Software Engineering for Develop Personal Website Using C#

ABSTRACT

The practical example personal website development (PWS) used in this chapter can be used by any user in the world for developing their website. Users will open PWS by type the address of the site using web browser. Users will navigate to website content, and if users have credential data, they can login to admin panel. PWA will install and configure on hosting support .net technology and Microsoft platform. Admin can add any new publication or book for electronic library section by login to admin site and admin can add new users and edit user profiles. After the authors introduce PWS system, they compare it with current content management systems that the user can use it from the internet. In the next sections, they analyse the system requirements functional and nonfunctional requirements. Each of the requirements will determine how to interact with other modules using uniform modeling language (UML) to describe objectives that PWS includes.

INTRODUCTION

This chapter describe five stages for personal website development (*PWS*), following section describe the analysis, design, implementation, testing and deployment and installation of the personal web site (PWS). This practical example proposed all features and procedures to develop the personal web site by introduce about personal web site system, analysis the user requirements, determine the functional requirements for objects for the system, the UML language will be used to analysis, and describe our web site to determine the details of objectives, scope, design model,

DOI: 10.4018/978-1-6684-4844-1.ch006

requirements and then we will study and monitoring, reporting mechanisms that will apply for developing the system.

There are many templates that already used to develop personal website system, such as the following:

Umbraco is a fully-featured, open source Content Management System loved by thousands for its flexibility and great editing experience. Use Umbraco for free and install, setup and host it yourself or let us take care of all of that for you on Umbraco Cloud.

SageFrame is a highly extensible Web Development Framework built on ASP. NET that helps you build dynamic web applications! We empower you with all the essential tools required to develop custom modules, applications, templates and various-purpose websites with ease and convenience.

But to use any of this system it add many complex to your site and many feature the user didn't use it but it already install to site, we need to create simple PWS using one page technology and connected to database (DB) with good speed and performance, the following points represent the functional requirement:

- Guest or user can access web site exam using computer or smart phone devices.
- eLibrary items will display to main page automatically after the the admin add it to admin section.
- admin will add ebooks, publication, user, edit user, edit ebooks from admin page connected to database.
- Responses by the candidates will be checked automatically.
- Age calculate for visitor by select data of birth the age in year, days, month display to user.
- Grade Calculator, user can add the number of Courses they have in a semester, then the system will generate at run time html table with label and textbox and calculate the average, GBA and use class student.
- Student can view the elibrary items directly in main page for download link for recourse student or user can download after login to site.

PWS is web application provide information for user about CV and all information who plane site administrator to publish about him related to experience, relevance, publications, electronic library and set of tools that the user can use it online, user will us the site include (student, guest and administrator) as describe in Figure 1, the main function for the system (System Functionalities) to create personal website, display for the user to read about personal and student or user can login to system for download books and other resources.

Figure 1. User requirements with functional requirements that will use by users

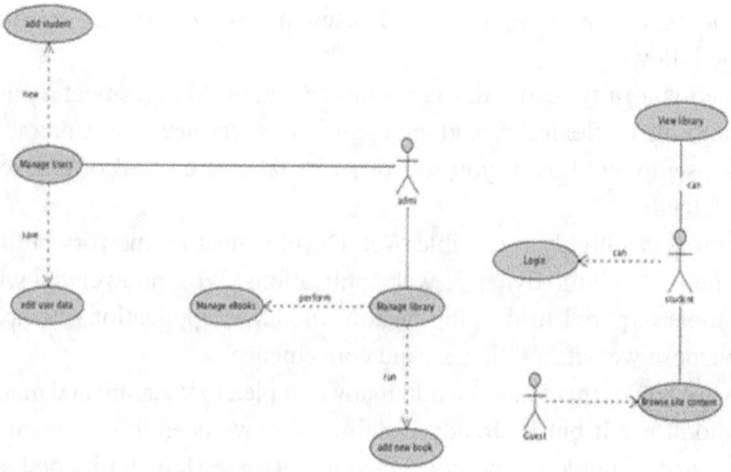

System User Requirements

User module:

1. can access eLibrary page.
2. Logging into the system.
3. view download ebooks links.
4. Appearing for the admin.
5. Can change the password.
   7. Resetting of forgotten password.
   8. use age and Grade calculator.

Administrator module

1. Back up for the database.
2. Full control to manage the system.
3. Create, delete, manage, update, view user account.
4. Editing, Deleting, Creating the eBooks.
5. Can change the password.

In this section, we try to analyze the overall activity of the PWS system by Use Case Diagram, Use Case Priority List, and Use Case Specifications. Actors are external entities that interact with the system, actor initiates system activities for the purpose of completing some tasks.

Actors for PWS are as follows: -

Administrator: Super user, adds user and manages system, do all admin task and access all system feature.

Student: User mostly a student who will login to download recourses.

user: any user who open the website.

Figure 2 outline the Login use case which describes a sequence of interactions between a user and system, without specifying the user interface, use cases describe the system functions that describe previous section, it describes the interactions between various user roles and the functionalities related to login and account management. The diagram features a central "User" actor, representing anyone interacting with the system. This "User" can take on several roles, represented by the actors branching out: "Admin," "Guest," and "student," they access the system using "Borrower", The "Admin," "student," and "Guest" have direct access to all four functionalities, it provides a clear visual representation of the user roles and functionalities associated with the login and account management aspects of the PWS system.

This use case encapsulates the functionalities related to login and account management, these functionalities are represented by oval shapes:

- Login, allows users to authenticate and access the system.
- Logout, enables users to securely exit the system.
- ChangePassword, allows users to change their password for security purposes.
- UpdateProfile, enables users to modify their profile information.

Figure 2. Use-case diagram for the PWS -LOGIN package

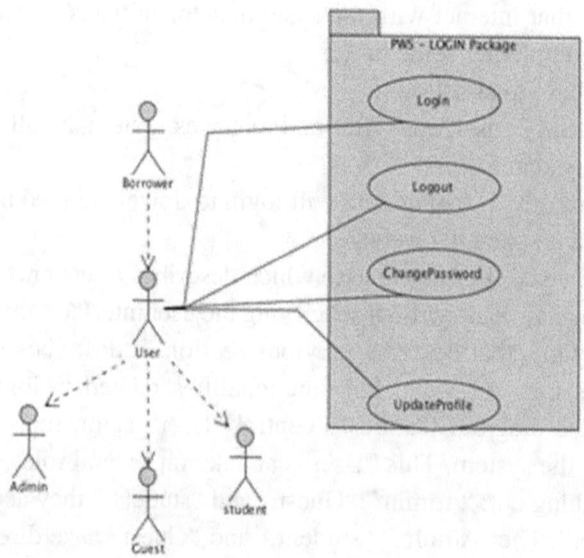

To describe user use case we need to detailing the use cases for "Login", "logout" and "change Password" within the PWS system to demonstrate the system's which focus on user authentication and security, this straightforward use case is initiated by an admin user who is already logged in and navigating the main menu, the user simply clicks on "logout," triggering to log them out, this ensures a secure exit from the system and has no alternative flows due to its simplicity, also admin can change a user's password. It assumes the admin is logged in and currently viewing user data, admin initiates the process by clicking "edit user data," which prompts the system to display an "edit button." then admin enters and confirms the new password before clicking "update," at which point the system successfully changes the password. However, there is a potential for an error if the admin enters an incorrect new password, In this case, the system displays an error message, prompting the admin to re-enter the correct old password. The logout function ensures users can securely exit, while the change password function, managed by admins, allows for password management, potentially preventing unauthorized access and enhancing security. Table 1 summarises all details with description for user use case.

Table 1. Users use case description

Use Case: Login Use Case	
ID	**PWS- LOGIN-01**
Brief Description	Login into the system.
Primary Actors	Administrator and student.
Secondary Actors	None
Preconditions	All authorized has the user name and password
Main Flow	1. All authorized user wants to login into the system. 2. User first must have username and password. 3. User enters their user name and password. 4. The system checks the user name and password. 5. The system accepts user name and password. 6. The use case ends
Post condition	The system accepts user name and password.
Alternative Flows	**action A:** if the users have no user name and password. 1. A.2 admin create new account and send to user. **action B:** if the entered user name and password is wrong. 1. B.3 Try with another user name and password.
Use Case: Login Use Case	
ID	**PWS-LOGIN-02**
Brief Description	The admin changes user Profile Information.
Primary Actors	Admin
Secondary Actors	Admin, user.
Preconditions	The Use-Case starts when: The admin is currently logged on the System. The admin is navigating on the main page. The admin click on edit user data on the main page.
Main Flow	The admin clicks 'Edit Profile Data". The System displays the 'users Page' and "update" Button at the top of the page. The User click on "update" add or fill profile fields then clicks 'Update". The System successfully changes the Profile. [ErrorProfileData]
Post condition	The Use-Case terminates when: • The User successfully update profile page.
Alternative Flows	[ErrorProfileData] • The User enters incomplete profile data '. • The System displays an error box telling the profile field error entered is incorrect.
Use case: logout	
ID	**PWS-LOGIN-03**
Brief Description	The User logs out of the System.
Primary Actors	Admin
Secondary Actors	Admin, user.

continued on following page

Table 1. Continued

Preconditions	The Use-Case starts when: • The User is currently logged on the System. • The User is navigating on the main menu.
Main Flow	1. The User clicks 'logout'. 2. The System logs the User out.
Post condition	The Use-Case terminates when: • The User successfully logout.
Alternative Flows	None
Use case: change Password	
ID	**PWS-LOGIN-04**
Brief Description	The admin need to changes user current password .
Primary Actors	Admin
Secondary Actors	Admin, user.
Preconditions	The Use-Case starts when: The admin is currently logged on the System. The admin is navigating on the edit use data.
Main Flow	The admin clicks 'edit use data'. The System displays the 'edit button. The admin enters new password and confirm new password then clicks 'update'. The System successfully changes the password. [Incorrect password]
Post condition	The Use-Case terminates when: • The admin successfully changes its password.
Alternative Flows	[Incorrect password] • The admin enters incorrect new password and clicks 'Change Password'. • The System displays an error box telling the User that the old password entered is incorrect.

Admin user case provides a visual overview of the admin's role in managing users, ebooks, and the overall system, it highlights the different functionalities available to the admin and how they interact with both users and the system's resources. Figure 3 outlining functionalities within an administrator package, likely for a library or digital resource management system, it showcases two main actors: the "Borrower," who utilizes the system to access resources, and the "Admin," responsible for managing the system, it divided into two packages: "User Package" and "System Package."

The "User Package" encompasses functionalities related to user management. "Manage User" allows the admin to oversee user accounts, while "EditUserDetails" enables the modification of user details. "Add eBook" and "Add user" functionalities allow the admin to add new ebooks and users to the system, "System Package" focuses on the administrative aspects of the system. "Library" allows the admin to manage the library's resources, and "add new ebook" it allows the admin to add new ebooks to the library. The "user" element, perhaps mislabeled, could represent user management within the system. "add new user" allows the admin to add new users to the system, and "manage Library" and "Manage users" enable the admin to oversee both the library's resources and its users.

Figure 3. Use-case diagram for the administrator package

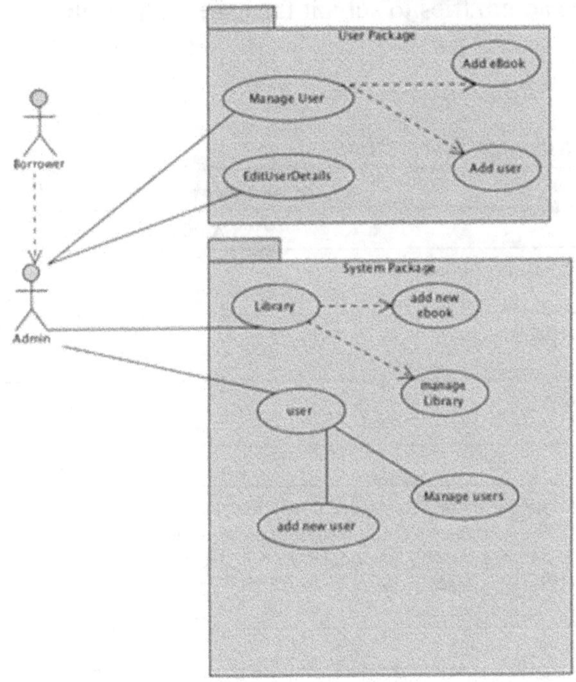

Above use case need to describe in details, for that each sub case describe in Table 2 with include Add eBook (PWS-Admin-01) use case which describes the process of an administrator adding a new eBook to the library, there are no preconditions, and the main flow involves the administrator wanting to add an eBook, adding it by name, and filling in the eBook data with an image, the use case ends with the eBook being successfully added. An alternative course of action exists if the eBook lacks an image; in this case, the administrator adds a default page cover.

Library Page (PWS-Admin-02) use case focuses on the automatic generation of the library table after the admin adds a book to the library, the library table is displayed on the admin page and the main page for users, the primary actor is the admin, with the user as a secondary actor. The main flow involves the admin logging in, accessing the add book page, filling in the book data, and viewing the generated library table. The use case ends with the library being generated. Alternative flows exist if the book data is wrong, prompting the system to check the data.

Edit-User-Details (PWS-Admin-03) use case details the admin's ability to modify user or eBook details, the preconditions state that the admin must be logged in and navigating the admin page. The main flow involves the admin clicking “Edit User data,” viewing the user data, selecting a user to edit, modifying the details, and submitting the changes, it ends with the user details being changed, an alternative flow exists if the admin tries to submit the page with invalid input, prompting an error message.

Table 2. Admin use case description

Use Case: Add eBook	
ID	**PWS-Admin-01**
Brief Description	Adding new eBook to the library.
Primary Actors	Administrator
Secondary Actors	none
Preconditions	
Main Flow	1. Administrator wants to add eBook in to library. 2. The Administrator adds the eBook by name and. 3. Administrator fill the eBook data with image. 4. eBook is added. 5. The use case ends.
Post condition	eBook is added.
Alternative Flows	**Alternative course of action A:** if the eBook has no image The administrator add default page cover.
Use Case: library page	
ID	**PWS-Admin-02**
Brief Description	Elibrary table will generate automatically after admin add book to library, library table will display on admin page after add or edit new item, and display in main page for user.
Primary Actors	admin.
Secondary Actors	User.
Preconditions	
Main Flow	1. Admin login to add book page 2. Admin fill book data. 3. After data add user show library table. 4. Library table page updated on main page. 5. library is generated. 6. The use case ends.

continued on following page

Table 2. Continued

Post condition	library is generated.
Alternative Flows	A: if the book data is wrong A.2 The system checks the data.
	Use Case: Edit-User-Details
ID	**PWS-ADMIN-03**
Brief Description	The Admin changes the user or ebooks details.
Primary Actors	Admin
Secondary Actors	None
Preconditions	The Use-Case starts when: • An Admin is currently logged on the system. • The Admin is navigating through the admin page.
Main Flow	1. The Admin clicks on 'Edit User data'. 2. The System displays the 'Edit User data'' menu. 3. The Admin click on table and select which user to edit. 4. The Admin edits the details he wishes to change and clicks the 'update' button. *[Invalid Input]* The System successfully changes the details and returns to admin menu.
Post condition	• The Use-Case terminates when: The User details have been changed.
Alternative Flows	[Invalid Input] • The Admin tries to submit the page with invalid input in one of the boxes. • The System displays an error message prompting the user to re-enter the user details. [Unable to update data]

NON-FUNCTIONAL REQUIREMENT

Performance: After completing the web site, the entire score of the web site calculated as the average time for page load in second in user browser.

Usability: The website should be user friendly and should require least effort to operate.

Portability: The website is develop using c#, HTML, CSS, JSP, etc. which are platform independent and can be transported to other servers with minimum effort.

Availability: • user can request site address at any time however can open site anytime to access the site information that include.

- administrator can create user on the admin panel at anytime.
- This PWS must run on multiple operating systems and support windows operating system.

Resolution: The PWS is characterized by accuracy of outputs provided by it is described accurately in terms of giving the outputs are correct and sound.

Reliability: The PWS is described as reliable in terms of errors and lack of errors that may affect its operation correctly or disable it for continuous periods of time.

Integrity: No person who is not authorized to enter this PWS where the system prevents anyone to enter the PWS only authorized persons through login page.

Reusability: The PWS can be reused as personal web site after update data and information that add.

SYSTEM DESIGN

System design is a crucial phase where you translate the abstract requirements of a system into a concrete blueprint for implementation, it involves defining the system's architecture, components, modules, interfaces, and data structures, laying the foundation for a robust, scalable, and maintainable software solution. A key aspect of system design is creating visual representations that communicate the system's structure and behavior, where UML diagrams, such as class diagrams.

The PWS Example have a relationships between these classes, the "administrator" class has a "Login" relationship with the "UserLogIn" class, indicating the administrator's ability to log in. Both "administrator" and "Library" classes have an "add book" relationship with the "Library" class, suggesting both administrators and the library itself can add books. The "Library" class has an "Add" relationship with the "LibraryItems" class, showing that the library contains library items. Lastly, the "administrator" class has an "Assign" relationship with the "userType" class, indicating the administrator's role in assigning user roles or managing user groups.

Figure 4 illustrates the system design for PWS, it utilizes UML notation to depict the classes, their attributes, methods, and relationships within this package. The "administrator" class represents the system administrator, encompassing attributes like ID, full name, username, password, type, and notes, it has access to various methods, including registration, eBook data editing, adding new books and users, user data editing, and password changes.

The "UserLogIn" class handles user authentication, with attributes like UserID, EventID, and Type, it includes methods for "Login" and "Logout."

The "Library" class represents the library functionality, with attributes for ID, Name, Description, Type, Time, Image, and Date. The library class provides methods for adding, editing, and deleting eBooks. The "LibraryItems" class represents individual items within the library, including attributes like ID, name, type, image, and methods for setting and retrieving question duration and correct answers. The "userType" class appears to represent different user roles within the system, with an attribute for user type and methods for managing user groups and accessing user lists.

Figure 4. Class diagram for AOES package

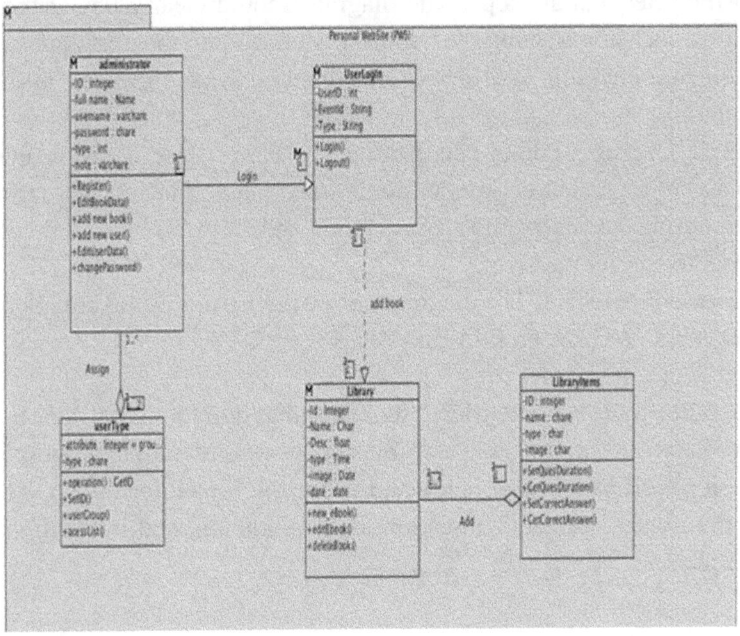

SEQUENCE DIAGRAM

System-level sequence diagrams in software engineering visually represent the interactions between different components of a system in a specific scenario, emphasizing the order of events and the messages exchanged. When designing a system for adding books to a library, a sequence diagram can clearly illustrate the steps involved and the interactions between the user (admin), the system, and the data store (library database). From our example you hav essentail step to draw a system-level sequence diagram as, follow:

- Identify Actors: The primary actor is the "admin" who initiates the process.
- Identify System Components: The key components include "Login," "Select Add New Book," "Library," and "Insert Book" it represent different functionalities or modules within the system.

- Define the Initial Action: The diagram starts with the admin entering their "UserName" and "Password" to log in.
- Show Message Exchanges: Use arrows to show the messages exchanged between the actor and the system components. For example, after the admin logs in, the system verifies their credentials ("Verify").
- Illustrate Sequential Steps: The diagram should clearly depict the order of actions, such as verifying the admin session, adding the eBook details, filling book data, clicking "add new," and finally, saving the book in the library database.
- Include Decision Points and Alternative Flows: Use decision nodes (diamonds) to represent choices or conditional paths. For instance, if there's an error during data entry, the process might end, or the system could prompt the admin to correct the error.
- Show the Final State: The diagram ends with the successful completion of the book addition ("eBook Edit End").

Figure 5 illustrates a system-level sequence diagram for adding an eBook by an admin, it effectively captures the step-by-step process of adding a book, including login, session verification, data entry, and saving the book information. It highlights the system's role in user authentication, data validation, and interaction with the library database.

Figure 5. System-level sequence diagram for add question and exam by teacher

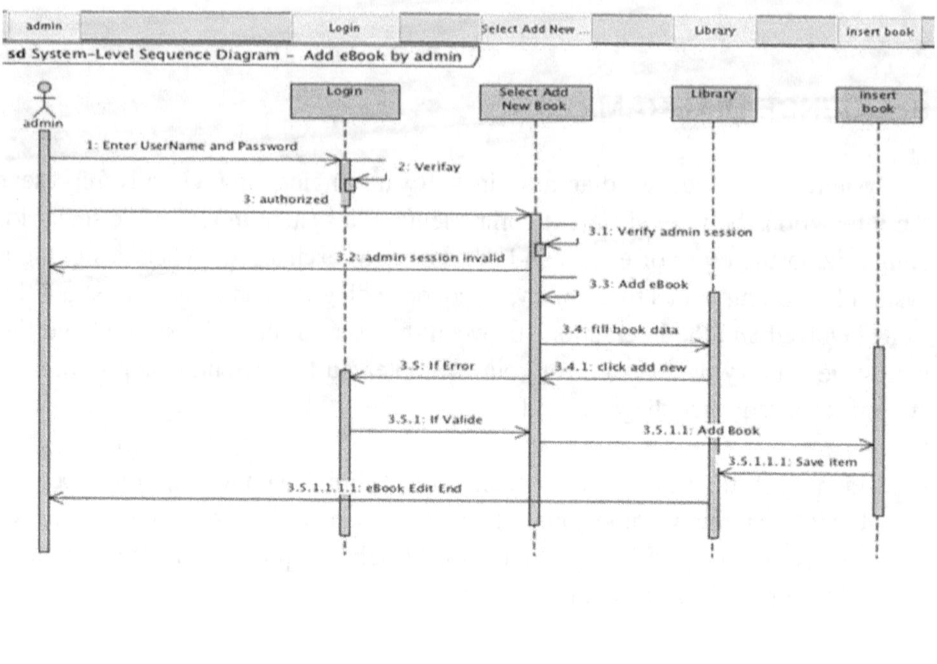

Finally, the user management package has dependencies on "Storage" and "Database" components, it effectively conveys the subsystem's modular structure, showcasing the separation of user management and admin functionalities, the hierarchical organization of packages, and the relationships between components which provides a clear and concise overview of the subsystem's architecture, promoting understanding and facilitating development and maintenance.

Figure 6 outlines the architecture of a personal website subsystem, highlighting its modular design and the dependencies between its components, it divided into two main packages: one for user management and another for the admin page, the user management package includes sub-packages for "Manage User," "Registered User," and "Student." The "Manage User" package further defines a "Guest User," while the "Registered User" and "Student" packages remain unspecified, likely representing areas for future development or more detailed diagrams.

Figure 6. Personal website sub system

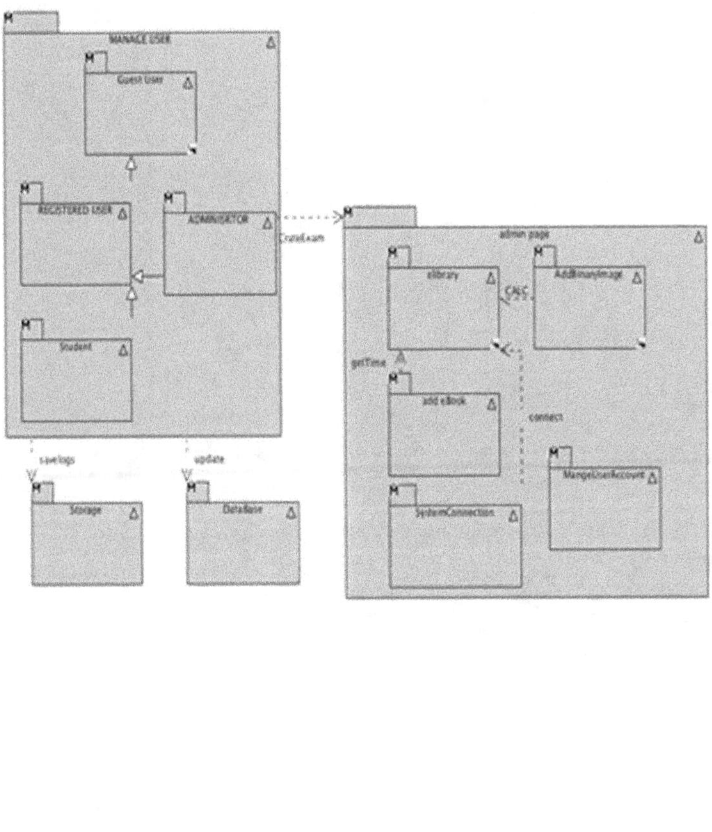

STATE DIAGRAM

State diagrams are a powerful way to visually represent the different states a system can be in and how it transitions between those states based on events or actions, it's important to map out the user's navigation path, outlining the main steps they can take within the system, as practical

Figure 7 illustrates a state diagram for the PWS system's user main navigation path, it showcases the different states a user can occupy, such as browsing the site, logging in, selecting a role, and performing actions like downloading books or editing data, it clearly depicts the transitions between these states and the events that trigger them.

Figure 7. User main navigation path for the PWS system

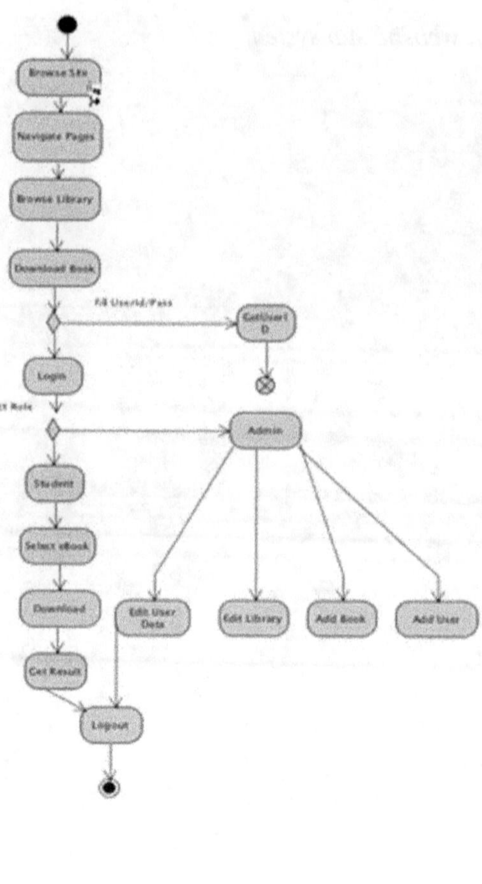

IMPLEMENTATION

implementation for PWS using visual studio for developing the page and used c# programing language, for data base using sql server 2015 [5,6]. Our example uses many sections such as menu, education, tools and publication menu, which include a very important data for visitor since it include research topics that the user normally search to contact with the teacher or site owner, Figure 8 shown Final result for the implement phase.

Figure 8. Publication menu

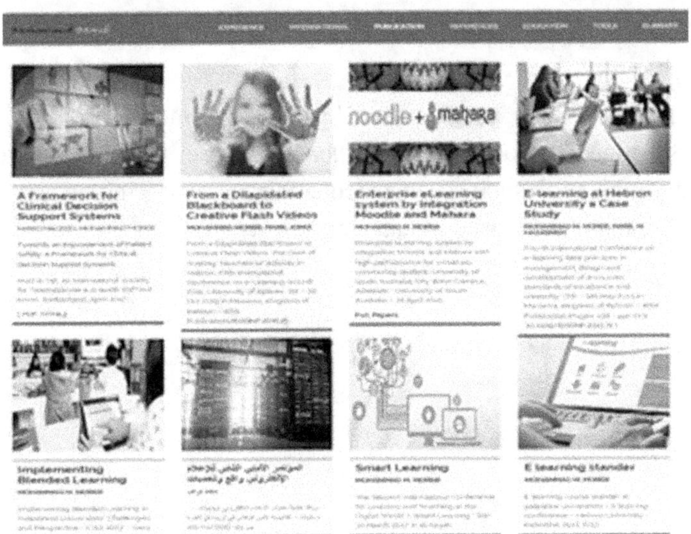

Tools menu shown in Figure 9, include set of online tools that develop and implement using c# language in PWS from analysis report guest user can select any of the following tools:

1. Age Calculator Online is c# page that calculate your age by select DOB from Calendar. This tools use class, if else statement, case and Query String.
2. Grade Calculator, user can add the number of Courses they have in a semester, then the system will generate at run time html table with label and textbox and calculate the average, GBA and use class student.
3. eLibrary include full pdf and publication that publish by user, as we descibe in use case after add book user need to reigiter to get the download link.

Figure 9. Age calculator menu

Each of the online tools' user can use it by click on demo link will display the page and allow user to use it. For download link redirect user to allow user to download source code for this tool, such as age calculator page which includes the main code to calc the age, main class used to calc age below:

```
protected void Button1_Click(object sender, EventArgs e)

{

Label1.Text = "";

Label2.Text = "";

Label3.Text = "";

DateTime birthDate = DateTime.Parse(TextBox1.Text);

int Days = (DateTime.Now.Year * 365 + DateTime.Now.DayOf-
Year) - (birthDate.Year * 365 + birthDate.DayOfYear);

int Years = Days / 365;

int month = (Days % 365) / 30;

if (Days >= 365)
```

```
Label1.Text = "Your age: in years = " + Years + " And " +
month.ToString() + " Months";

Label2.Text = "Your age in Days = " + Days;

Label3.Text = "Your age in Hours = " + Days * 24;

if (Days <= 0)

{

Label1.Text = "Please Add Correct Date Of Bairth !!!! ";

Label2.Text = "";

Label3.Text = "";

}
```

GRADE CALCULATOR

In grade page by default user will add two marks then click on calc to calc the average for the course and GPA as describe in previous use case, if the courses that the user will add more than two course user will select how many course who will add max is 9 courses, then after select number of courses at run time using c# code the page will create html table each row have label control and textbox control, all of this generate at run time, for calc avg using class to reduce the code in the page request, next screen and code describe the main process for calc avg page as shown in Figure 10.

Figure 10. Grade calculator

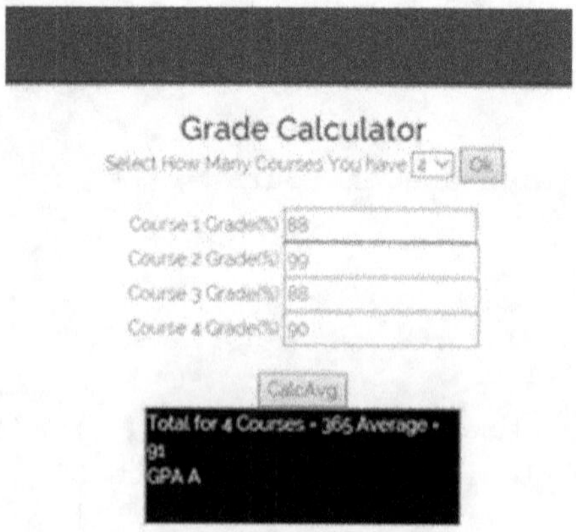

Below main Class use for calc grade is student Class which represents a student, with properties for name and average grade ("avg"). The "avg" property has getter and setter methods to ensure the average value is within the valid range (0-100).

```
protected void Button1_Click(object sender, EventArgs e)

{

int total = 0;

string tvalue = string.Empty;

string controlid = string.Empty;

for (int i = 0; i <= form1.Controls.Count; i++)

{

controlid = "textBox" + i.ToString();

TextBox tb = form1.FindControl(controlid) as TextBox;

if (tb != null)


{
```

```
// Response.Write(tb.Text.ToString());

total += int.Parse(tb.Text.ToString()) ;

}

int coun = int.Parse(DropDownList1.SelectedValue);

int avg = total / coun;

student std1 = new student("Student", avg);

if (total >= 40)


xxx.Text = "Total for " + coun + " Courses = " + total + "
Average = " + avg + "\n" + "GPA " + std1.AvgLeeter ;

else

xxx.Text = "Error Please Add Correct Value ...!!!!" ;

}

}

protected void Page_Load(object sender, EventArgs e)

{

string DOB = Request.QueryString["firstname"];

if (DOB == "12/3/1981")


xxx.Text = "Happy Birth Day, Just to reminde you ..... " +
DOB;

int coun = int.Parse(DropDownList1.SelectedValue);
```

```
int rowCnt = coun;

for (int i = 1; i <= coun; ++i)

{

TableRow tRow = new TableRow();

TableCell tCell = new TableCell();

TextBox txt = new TextBox();

txt.ID = "textbox" + i.ToString();

txt.Attributes.Add("runat", "Server");

txt.Text = "0";

Label lbl = new Label();

lbl.ID = "labelC" + i;

lbl.Text = " Course " + i + " Grade(%) ";

Table1.Rows.Add(tRow);

tRow.Cells.Add(tCell);

tCell.Controls.Add(lbl);

Table1.Rows.Add(tRow);

tRow.Cells.Add(tCell);

tCell.Controls.Add(txt);

}

public class student
```

```
{
public string name { get; set; }
private int v1 = 0;
public student(string StdName, int average)
{
name = StdName;
v1 = average;
}
public int avg
{
get
{
return v1;
}
set
{
if (value > 0)
v1 = value;
else if (value <= 100)
v1 = value;
```

```
}
}
public string AvgLeeter
{
get
{
string AvgLeeter = string.Empty;
if (avg >= 90)
AvgLeeter = "A";
else if (avg >= 80)
AvgLeeter = "B";
else if (avg >= 70)
AvgLeeter = "C";
else if (avg >= 60)
AvgLeeter = "D";
else
AvgLeeter = "F";
return AvgLeeter;
}
}
```

```
public string AvgLeeterC

{

get

{

string l;

switch (avg)

{

case 90:

l = "A";

break;

case 80:

l = "B";

break;

case 70:

l = "c";

break;

default:

l = "F";

break;

}
```

```
return 1; }
```

ELIBRARY PAGE

This page enable admin to add new recourses such as ebooks, document any electronic resources to main page by login to admin page then click on add eBook and add the data in table, after admin save the eLibrary will automatically updated in main page, this phase depending on the previous use case as shown in Figure 3 and Table 2.
Main implementation code used in this page as following:

- Contention with data set.
- Data grid with columns.
- Read binary image from database.
- User query string to hander the image id request from data grid, to reduce code in main page used Handler1.ashx that include code to add to image column in data grid.

```
<asp:Image ID="Image1" runat="server" ImageUrl='<%#
"Handler1.ashx?id_Image="+ Eval("id") %>' Height="400px"
Width="600px"/>
```

Figure 11. eLibrary page

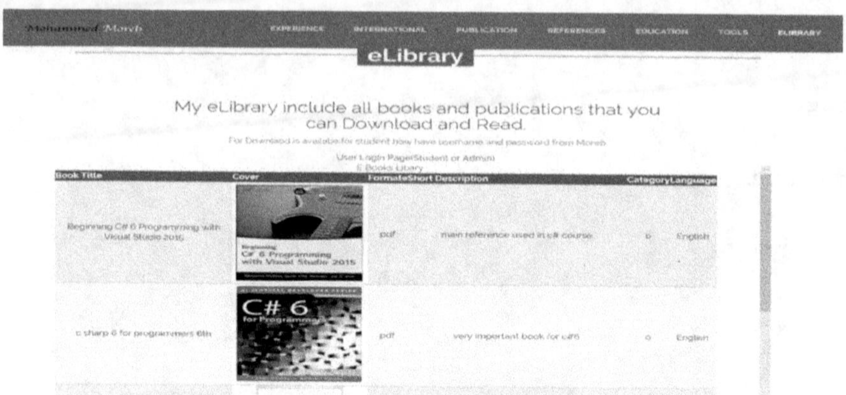

Main class - Handler1.ashx

```
public void ProcessRequest(HttpContext context)

{

string query, constr;

SqlConnection con;


string displayimgid=context.Request.QueryString["id_Image"].
ToString();


constr = ConfigurationManager.ConnectionStrings["mycon"].
ToString();

con = new SqlConnection(constr);

con.Open();

query="select img from ebooks where id="+displayimgid;

SqlCommand com=new SqlCommand(query,con);

SqlDataReader dr=com.ExecuteReader();

dr.Read();

context.Response.BinaryWrite((Byte[])dr[0]);

context.Response.End(); }
```

Admin can login as shown in Figure 12 to edit the system by click in login page, after click on login page will add user name and password if successfully then main admin page will display as next screen.

Main implementation code used in this page as following:

- Session used by other page to determine admin session key.
- Contention with data reader.

- Validation control (email, require).

Figure 12. Admin page login and admin dashboard

For login code after user add username as email and password, the system will cheek if credential data is valid the system will determine if the user is administrator will redirect to admin page if user is student will redirect to student portal allow them to download the eBooks, next code show the main steps to logon users.

```
SqlConnection con = new SqlConnection(ConfigurationManager.
ConnectionStrings["mycon"].ToString());

protected void Button1_Click(object sender, EventArgs e)

{

string uid = TextBox1.Text.ToString();

string pass = TextBox2.Text.ToString();

con.Open();

string qry = "select * from userLogin where UserName='" +
uid + "' and pass='" + pass + "'";

SqlCommand cmd = new SqlCommand(qry, con);

SqlDataReader sdr = cmd.ExecuteReader();
```

```
if (sdr.Read())

{

Session["LoginTime"] = DateTime.Now.ToString();

Session["sessionkey"] = TextBox1.Text;

Session["usertype"] = sdr.GetValue(4).ToString() ;

Label4.Text = sdr.GetValue(4).ToString();

if (sdr["userType"].ToString().Trim() == "100")

Response.Redirect("adminMain.aspx");

if (sdr["userType"].ToString().Trim() == "0")

Response.Redirect("~/student/Default.aspx");

}

else

{

Label4.Text = "UserId & Password Is not correct Try
again..!!";

}

con.Close();
```

Main implementation code used in add New eBook page as following:

- Insert sql command with parameter and Contention with data set.
- Data grid with columns.
- Validation for user input.
- Read binary image from database.

- User query string to hander the image id request from data grid, to reduce code in main page used Handler1.ashx that include code to add to image column in data grid, Figure 13 shown add new book form.

Figure 13. Add new book

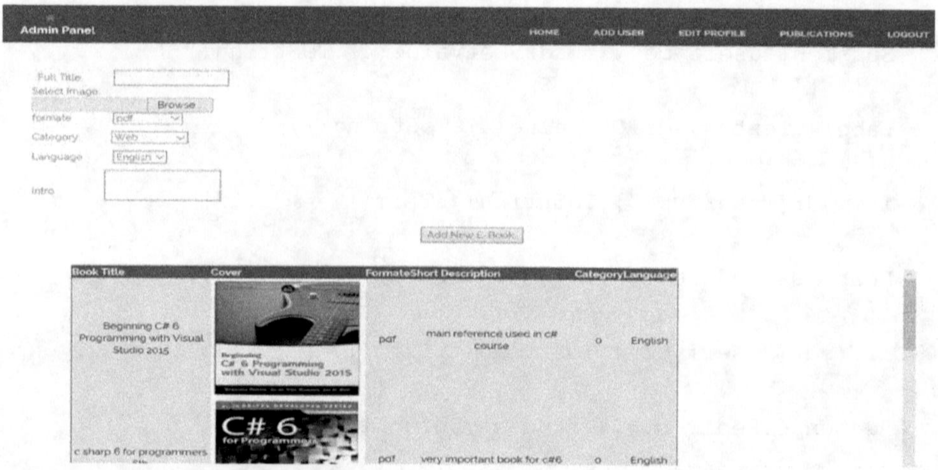

Code used to add image and save to image table.

```
private void Imageupload()

{

if (FileUpload1.HasFile)

{

int imagefilelenth = FileUpload1.PostedFile.ContentLength;

byte[] imgarray = new byte[imagefilelenth];

HttpPostedFile image = FileUpload1.PostedFile;

image.InputStream.Read(imgarray, 0, imagefilelenth);

connection();
```

```
   query = "Insert into ebooks (title,formate,intro,img,cate-
gory,language) values (@title,@formate,@intro,@img,@category,@
language)";

   SqlCommand com = new SqlCommand(query, con);

   com.Parameters.AddWithValue("@title", SqlDbType.VarChar).
Value = TextBox1.Text.Trim();

   com.Parameters.AddWithValue("@formate", SqlDbType.VarChar).
Value = title.SelectedValue.ToString();

   com.Parameters.AddWithValue("@intro", SqlDbType.VarChar).
Value = notetxt.Text;

   com.Parameters.AddWithValue("@category", SqlDbType.VarChar).
Value = category.SelectedValue.ToString();

   com.Parameters.AddWithValue("@language", SqlDbType.VarChar).
Value = language.SelectedValue.ToString();

   com.Parameters.AddWithValue("@img", SqlDbType.Image).Value =
imgarray;

   com.ExecuteNonQuery();

   Label1.Visible = true;

   Label1.Text = "E Book add successfully";

   imagebindGrid();}}

   public void imagebindGrid()

   {
```

```
   connection();


   query = "Select id,title,formate,intro,img,category,language
from ebooks";

   SqlCommand com = new SqlCommand(query, con);

   SqlDataReader dr = com.ExecuteReader();

   Gridview1.DataSource = dr;

   Gridview1.DataBind();

   }
```

Main implementation code used in Edit Library eBook as following:

- Data grid with columns.
- Edit, delete data grid column.
- Insert data sql command;
- Using admin master page, Figure 14 shown Edit Library ebook.

Figure 14. Edit library ebook

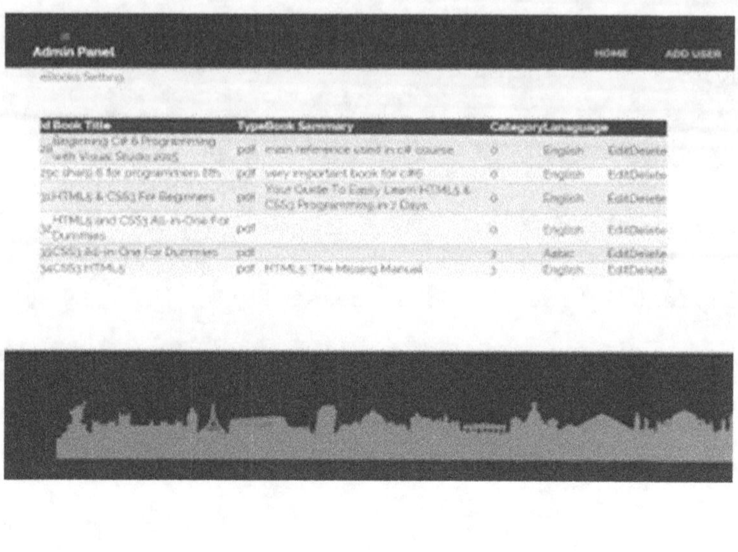

Delete data grid code:

```
protected void GridView1_RowDeleting(object sender, Grid-
ViewDeleteEventArgs e)

{

GridViewRow row = (GridViewRow)GridView1.Rows[e.RowIndex];

Label lbldeleteid = (Label)row.FindControl("lblID");

conn.Open();


SqlCommand cmd = new SqlCommand("delete FROM ebooks where
id='" + Convert.ToInt32(GridView1.DataKeys[e.RowIndex].Value.
ToString()) + "'", conn);

cmd.ExecuteNonQuery();

conn.Close();

LoginBind();

}
```

Update data grid code:

```
protected void GridView1_RowUpdating(object sender, Grid-
ViewUpdateEventArgs e)

{


int userid = Convert.ToInt32(GridView1.DataKeys[e.RowIndex].
Value.ToString());

GridViewRow row = (GridViewRow)GridView1.Rows[e.RowIndex];

Label lblID = (Label)row.FindControl("lblID");
```

```
TextBox text0 = (TextBox)row.Cells[0].Controls[0];

TextBox text1 = (TextBox)row.Cells[1].Controls[0];

TextBox text2 = (TextBox)row.Cells[2].Controls[0];

TextBox text3 = (TextBox)row.Cells[3].Controls[0];

TextBox text4 = (TextBox)row.Cells[4].Controls[0];

TextBox text5 = (TextBox)row.Cells[5].Controls[0];

GridView1.EditIndex = -1;

conn.Open();


SqlCommand cmd = new SqlCommand("update ebooks set title='"
+ text1.Text + "',formate='" + text2.Text + "',intro = '" +
text3.Text + "',category = '" + text4.Text + "',language = '" +
text5.Text + "'where id='" + userid + "'", conn);

cmd.ExecuteNonQuery();

conn.Close();

LoginBind();

}
```

Main implementation code used in Add New User Page as following:

- ContentPlaceHolder to handle html table .
- Display user data in grid column.
- Insert data sql command;
- Using admin master page.
- Validation control to validate user input(require, email and compare validation).

```
protected void upload(object sender, EventArgs e)

{

SqlCommand insert = new SqlCommand("insert into UserLog-
in(userName, pass,usertype,fullname,note) values(@userName, @
pass, @userType, @fullName,@note)", conn);

insert.Parameters.AddWithValue("@userName", username.Text.
Trim());

insert.Parameters.AddWithValue("@pass", pass.Text.Trim());

insert.Parameters.AddWithValue("@usertype", type.Selected-
Value.ToString());

insert.Parameters.AddWithValue("@fullName", fullname.Text.
Trim());

insert.Parameters.AddWithValue("@note", note.Text.Trim());

try

{

conn.Open();

insert.ExecuteNonQuery();

lblmsg.Text = "Register done ! Record Inserted Succesfully
into the Database";

fillGrid(); //dispaly data after add it
```

```
}

catch (Exception ex)

{

lblmsg.Text = "Error: " + ex.Message;

}

finally

{

conn.Close();

}

}
```

Main implementation code used in Student Main Page as following:

- ContentPlaceHolder to handle html datagride .
- Display eLibrary data in grid column.
- Display ebook links in grid column;
- Using admin master page, Figure 15 shown student main page.

Figure 15. Student main page

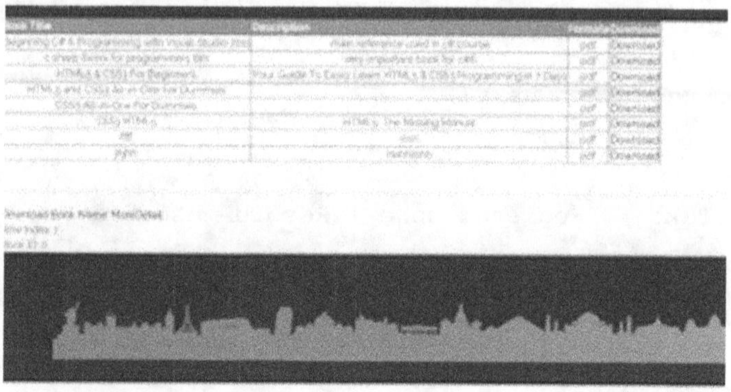

Main Class used to display data for student page:

```
protected void Page_Load(object sender, EventArgs e)

{

if (!IsPostBack)

{

string consString = ConfigurationManager.Connection-
Strings["mycon"].ConnectionString;

using (SqlConnection conn = new SqlConnection(consString))

{

SqlCommand cmd = new SqlCommand("SELECT * FROM [ebooks]",
conn);

cmd.CommandType = CommandType.Text;

conn.Open();

SqlDataReader drStudents = cmd.ExecuteReader();

GridView1.DataSource = drStudents;

GridView1.DataBind();

}

}

}
```

```
protected void GridView1_RowCommand(object sender, GridView-
CommandEventArgs e)

{


Int32 nRowIndex = Convert.ToInt32(e.CommandArgument.To-
String());


Int32 nProductID = Convert.ToInt32(GridView1.DataKeys[n-
RowIndex].Value);

HyperLink1.Text = "Download Book Name: " + e.CommandName;

Label2.Text = "Row Index: " + nRowIndex;

Label3.Text = "Book ID: " + nProductID;

}
```

TESTING

We have many testing techniques such as: Unit Testing, which test individual units of code (classes, methods) in isolation to ensure they function correctly. Integration Testing by Test how different components of the system interact with each other. System Testing by test the system as a whole, simulating real user scenarios. User Acceptance Testing (UAT) which Involve actual users in the testing process to gather feedback and identify usability issues.

Testing phase it's important as you're taking to testing your system is important for ensuring its quality and reliability. Focusing on user input validation is particularly crucial to prevent errors, security vulnerabilities, and unexpected behavior, below we breakdown of our example based on the provided information and suggestions for improvement:

- Login Page Validation, we should correctly identified three essential validations for the login page:

1. Required Username: Ensuring that the user cannot submit the form without entering a username. This can be implemented using a "RequiredFieldValidator" control in ASP.NET as shown in Figure 16.
2. Required Password: Similar to username, the password field should also be mandatory.
3. Email Format: we Verifying that the entered username conforms to a valid email format ("RegularExpressionValidator") can be used to achieve this, matching the input against a regular expression for email patterns. In login page as shown in Figure 16 we use three validations to manage user input as following:

    a. Require user name
    b. Require password
    c. Username should be as email.

- Expanding Test Cases, while these three validations are a good start, consider expanding our test cases to include:

    1. Invalid Username/Password Combinations, by test with incorrect usernames and passwords to ensure the system handles authentication failures correctly and displays appropriate error messages.
    2. Boundary Conditions, by test with usernames and passwords that are at the minimum and maximum allowed lengths. This helps identify potential issues with data truncation or validation logic.
    3. Special Characters and SQL Injection, by test with usernames and passwords containing special characters, spaces, and even potential SQL injection attempts (e.g., '; DROP TABLE users;'). This ensures that the system sanitizes inputs and prevents security vulnerabilities.
    4. Performance by measure the time it takes for the system to validate user credentials and provide feedback.

- Beyond the Login Page, all user inputs throughout the system should be thoroughly tested, pay close attention to areas where data is manipulated or stored, such as exam creation, question management, and result calculation.

Figure 16. Testing

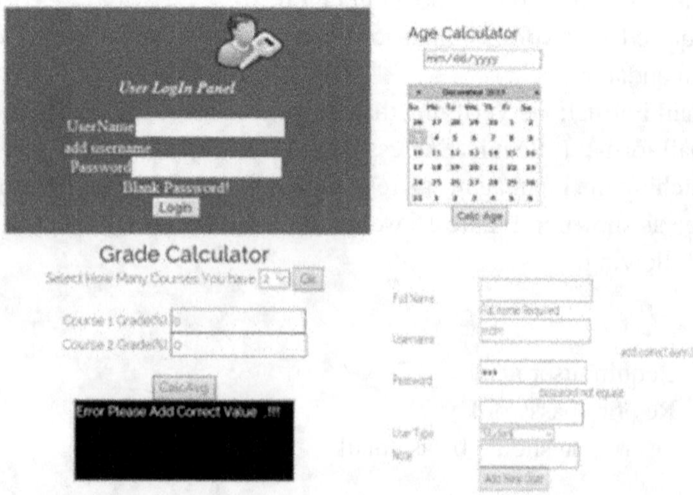

RESULTS

This chapter detailed the development of a personal website (PWS) using C# and ASP.NET, demonstrating the practical application of software engineering principles. We successfully implemented the core functionalities outlined in the analysis phase, including user authentication, eBook management, grad calc, and an administrative interface. By leveraging UML diagrams like use case diagrams, class diagrams, sequence diagrams, state diagrams, and data flow diagrams, we created a clear and comprehensive design for the PWS, we then translated this design into a functional website, complete with user interfaces, data storage, and backend logic.

CONCLUSION

This practical example showcased the power of using a structured software engineering approach for web development, by following the key phases of analysis, design, implementation, and testing, we created a robust, functional, and user-friendly PWS. This example serves as a valuable guide for readers interested in developing their own websites or web applications. The step-by-step explanations, code snippets, and visualizations provided in this chapter empower readers to apply these principles to their projects, promoting a deeper understanding of the software development lifecycle and enabling them to create high-quality software solutions.

Chapter 7
Implementing Point of Sale Web Application System Using php

ABSTRACT

This chapter outlines a comprehensive software engineering approach to developing a point-of-sale (POS) system suitable for any supermarket. The authors begin by examining the current system, analyzing its requirements, and identifying the functional needs of system objects. Utilizing the unified modeling language (UML) for analysis and system description, they meticulously define the project's objectives, scope, design model, and requirements. The chapter then delves into monitoring and reporting mechanisms crucial for the system's development. Throughout this process, the visual paradigm tool serves as the primary software design tool, facilitating the creation of visual models and diagrams.

INTRODUCTION

POSWA is a point of sale (POS) web application designed for responsive devises, supports receipt print, barcode scanner, Sales Report, Customers, Products and Sales page that enable casher to offer the products for sale either using cash method or add to customer account as a debit method. POSWA can be used by any supermarket and several companies specialized in buying wholesale to retail sales.

DOI: 10.4018/978-1-6684-4844-1.ch007

POSWA can also be used by various products and services suppliers like clothing shops, accessories, furniture, electronics, perfumes, pharmacy and many others.

POSW needs a PHP server compatible to run. The system can be used by installing it locally or online on a remote server. The system will work for any operating system and user can access the system application from a desktop computer or a mobile devise. In Summary as we introduce in Chapter 2 the System Life Cycle as:

1. Requirement Specification

The first step in POSWA project was to collect all the information about the existing system and try to understand the problems that it facing and to find a solution for it.
2 Analyzing

After the requirement specification step, we start the analyzing phase and, in this step, we tried to describe our new step by using some UML's diagrams, which included use case diagrams and class and interface diagrams.

3. Design

After the analysis phase, we start the design step in this step also we used in details UML's diagrams for each case that determine in previous phase, Design is the abstraction of a solution and view patterns seen in the analysis phase to be a pattern in a design phase, which included state diagrams and sequence and activity and some other diagrams that represent the interaction between components and represent stream functioning of the system.

4. Implantation

The final stage in our project is the implementation step where we have assimilated all classes to the programming language. There have been difficulties especially with the reservation part because we need to consider all the dates that starting from the reservation date until the return date also was a bit difficult to compare the dates, but we have tried to implement as much as possible. The system has been developed using MySQL database and PHP by using macromedia Dreamweaver to write code and design html page and user interface and our database (this report documents all details for this stage).

BACKGROUND

Point of sale systems are used at checkout totally items for payment, to process a payment, and to update a store's inventory. These systems have computing capabilities and often use multiple data-collection devices. These data-collection devices are typically added as peripherals to a host-computing device. In other words, the point of sale (POS) system is the place where your customer executes the payment for goods or services bought from your company.

Most of researches now days' work on intelligent POS that include information technology system taking the form of a management console, central server, point of sale system, software running on a point of sale system and software running on a customer's mobile phone. The software on the customer's mobile phone lets a customer interact with the point-of-sale system, allowing the customer to view a pending economic transaction, optionally add or remove items to the transaction, and pay for the transaction using payment information stored on the mobile phone.

Primarily, the scope pertains to the POSWA features for serving a supermarket such as Halhul Mall. It focuses on the supermarket, the stakeholders and applications, which allow for using and accessing POS web application from a computer devise or a smart phone. This POSWA is also aimed at specifying requirements of software to be developed but it can also be applied to assist in market or supermarket. The system can be used for small businesses they have a related user requirement's as POSWA had like most of modern and intelligent POS requirements needed by user in now days comparing this with many POS systems.
Features:

- Inventory management
- Sales Management with two types of payments (cash, debit).
- Printing Received.
- Printing items labels
- Purchases Invoices
- Daily report
- Monthly report

In short, the purpose of this analysis report document is to provide a detailed overview of POSWA software, its parameters and goals. This document describes the POSWA target audience and its user interface, hardware and software requirements. It defines how do our client, team and audience see the product and its functionality. Nonetheless, it helps the team designer and developer to implement the POSWA software delivery lifecycle processes. Some of these POS System developed and used in the market.

Arabicsw it desktop application: This system works on one machine, and one user can use the program, the program contains many accounting and management reports, enabling the administrator to make sound decisions based on the resulting reports. The report builder can also be used to produce various reports that may be required by the work. full control of bank transactions and checks, where you can control incoming or outgoing checks easily. ShopKeep is POS system for retailers, it's an easy-to-use, cloud-based service that offers competitive pricing, doesn't require a lengthy contract, and has tons of features help to run and grow your retail business. Openbravo is a company that specializes in open-source ERP. It started as an university management project during the mid-1990, but later turned into a company known as Tecnicia in 2001, and today known as Openbravo. In 2006, Openbravo received $6.4 million in venture capital funding. In 2008, it received an additional $12.5 million investment fund. With the infusion of money, the company has aggressively expanded its reach to the potential ERP buyers around the world. It estimates a significant increase in revenues as more users purchase and adopt its software. The size of the company has grown from 70 employees in 2008 to 100+ employees and management team today.

To determine the Statement of problems in the current system we find the best POS systems for small businesses, we asked small business owners about which systems they use and what do they like and dislike about them, and what are the issues they have in current system for the users of POS system. We also red customer reviews to see what users with firsthand experience think about them. Figure 1 shows the top three most popular solutions used. This system includes all the functions that the market needs, but to use it, high professionalism is needed in terms of skills and training. In some cases, even accounting specialists in business is required.

Figure 1. Top 3 most popular solutions

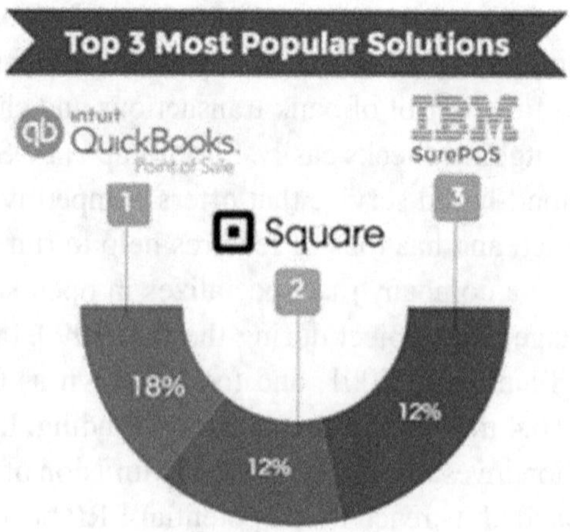

We then compiled a comprehensive list of POS systems from several sources including open-source system and licensed system, including customer research, review sites and business websites, those we knew about or that were previously pitched to us, and some from our previous experience in this fields for small businesses. Finally, we have more than 10 POS systems to choose from.

Next, we researched these vendors and narrowed down our list based on use cases to support Arabic language and use the system to fit our user requirements to be simple and support the Halhul Mall case, as if we apply to this small business then we will use the system for all the markets in the same case requirement's. Our methodology includes browsing each vendor's website, trying out demos, watching video tutorials and reviewing how-to guides, installation manuals, articles and other available resources. Additionally, in some cases we contacted vendors and asked them several questions as small business owners in the market for a new POS system.

After much consideration, our short list included Arabicsw, ShopKeep, Openbravo, Bindo, and TouchBistro. Following are the criteria we used to evaluate each POS system:

- Cost, such as yearly fees.
- Contracts and service terms
- Ease of setup and use
- Back-end features, such as reporting and inventory management
- Customization options
- Third-party integrations
- Customer service

- Inventory management
- Sales Management with two type of payments (cash, debit).
- Printing Received.
- Printing items labels
- Purchases Invoices
- Daily report
- Monthly report

Figure 2 shows the most important factors in POS software. When to purchase is ease to use and functionality, from this we can determine that the complexity of the current system have answered the question about using paper based in our case and user didn't use current system to manage customers and suppliers.

Figure 2. Important factor in POS

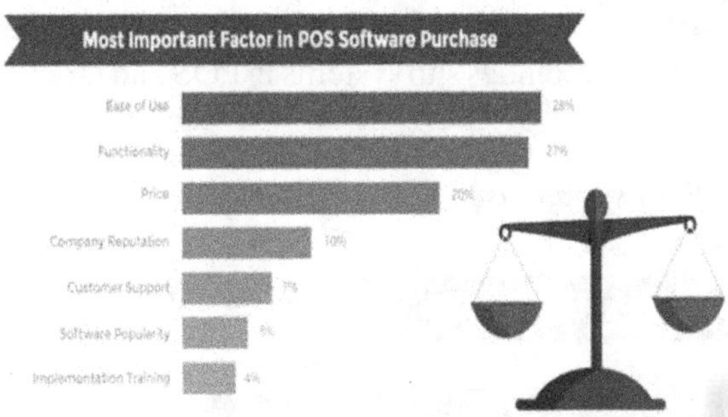

Figure 3 shows deployment method in POS. Current today is web-based or installed. In current system, the system is installed on local computer and can be used from one casher, after analysis the required deployment method they want, the users answer was that they need to install locally on local server and be able to run from computer and smart phone, so new system will support two deployment methods we have.

Figure 3. Deployment method

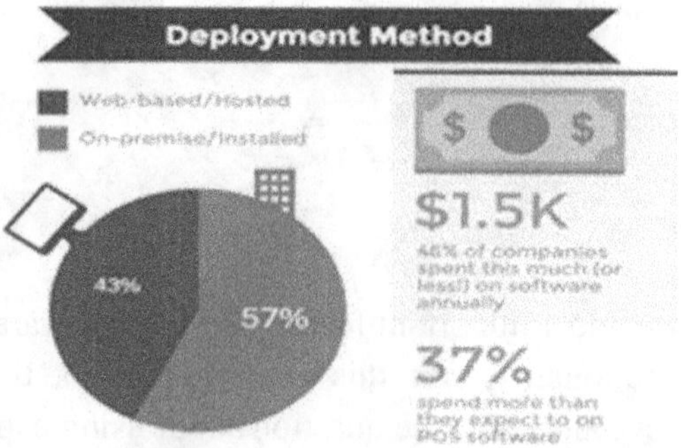

Figure 4 shows a summarize for most of POS method used features that the user require and need to use, by comparing these features with current system, it is showed that the current system doesn't support what the user want, and most of the feature that Al-Arabi POS includes sub systems in POS and ignores the POS features.

Figure 4. Method POS system used features

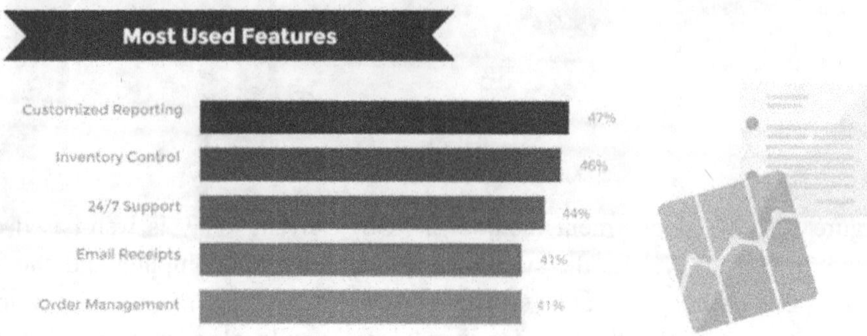

All the payment and purchasing process doing by paper based and the current system used as calculator for products that the user purchase, for suppliers should add in the system and manage by current system but after interview the manager at Halhul mall they used paper based and didn't use the already system related the complexity and have many steps to add the products and need accounting specialist to use the system, this needs a lot of time, the whole process of add products to inventory calculate suppliers summation and manage customer debit was done manually until date. It is very time wasting, also it is difficult to keep the customer's data and supplier's management in this case, as an instance if the Halhul mall case each day casher at the end of the day need 2 hours to fill information for customer and supplier's data in papers * 30 days in month = 60 hours, for supplier's report need 3 hours to prepare it each week, so each month need about more than 12 working days to manage the records manually using paper base as we use al-Arabi POS system, it's worthily mention that this system used widely in Palestine and have more than 950 registered license, and most of the clients have the same problem for our case.

New system Point of sale system (POSWA) can be used by any supermarket or small business for developing their strategy from paper base to electronic system and reduce time in managing the customers and suppliers.

From 5 show, main reasons for switching from Al-Arabi POS (current POS Software) used by Halhul mall.

Figure 5. Reasons for switching from previous POS software

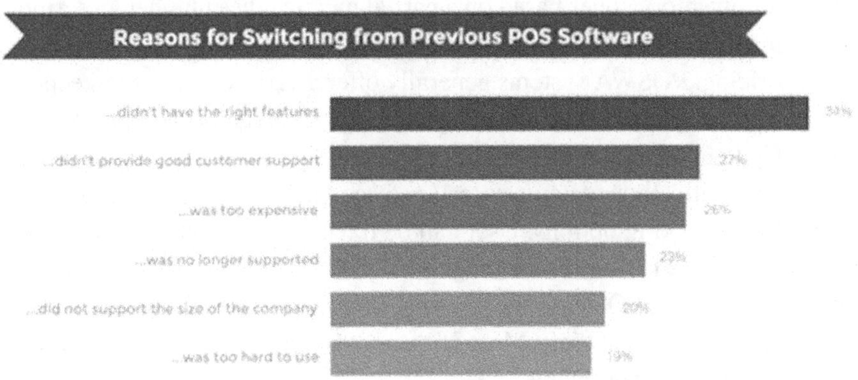

POSWA software can summarize in simple steps for administrator, Seller and buyer as users will open the it, this system will install and configure on local computer or can be run from online server, after casher login to the system he/she can open POSWA and POS screen and start to add products for customer after finish of shopping cart the casher will add the amount of cash then print Receipt for customer, system user can add product for the system or update existing info and price for products, POSWA can view reports for date range include profit amount for selected date.

POSWA will work to interact and get support other technology such as barcode scanners, cash drawers, and receipt printers can be interfaced within the POSWA software to create a powerful point-of-sale solution. Product and bin labels containing barcodes can be printed directly from the POSWA software.

You can analyze sales data, figure out how well all the items on your shelves sell, and adjust purchasing levels accordingly.

You can maintain a sales history to help adjust your buying decisions for seasonal purchasing trends.

You can improve pricing accuracy by integrating bar-code scanners and credit card authorization ability with the POS system.

POSWA system Features include the following:

- Ease of use. Look for software with a user-friendly graphical interface and can run from mobile devise beside computer desktop.
- Entry of sales information. Most systems allow you to enter inventory codes either manually or automatically via a bar-code scanner. Once the inventory code is entered, the systems call up the standard or sales price, compute the price at multiple quantities and provide a running total. User can easy enter sales manually when needed by letting you search for inventory codes based on a partial merchandise number, description, manufacturing code or vendor.
- Pricing. POSWA systems generally offer a variety of ways to keep track of pricing, including add-on amounts, percentage of cost, profit.
- Updating product information. Once a sale is entered, these systems automatically update inventory and accounts receivable records.
- Sales tracking options. User can be tracking sales report for any day or date range by select date range.
- Security. In POSWA, it's important to keep tight control over cash receipts to prevent theft. Most of these systems provide audit trails so you can trace any problems.
- Taxes. When the user add, product can add the taxes type and it will add automatically to selling price.

FUNCTIONAL REQUIREMENTS

This subsection contains the requirements for the POSWA. These requirements are added after comparing current system function that include and what the use require from the system to run in simple and intelligent, then these requirements refined into use case diagrams in next document and to sequence diagram to best capture the functional requirements of the system.

1.1. Product Categorizations

- The system shall enable user to add categorization to include product.

1.2. Products screen

- The system shall allow manager to add new product assign to product category.
- The system shall allow manager to edit current product.
- The system shall allow manager to search for products.
- The system shall display available products to the casher on POS screen and user casher can select promotions and add to customer invoice.
- The system shall calculate tax for the products.
- The system shall display tax information for the invoice.

1.3. Provide Search facility.

- The system shall enable casher to enter the search using product bar-code or name.
- The system shall display the matching product based on the search and add automatically to bill.
- The system shall notify the user when no matching product is found on the POS screen.

1.4. Customer screen.

- The system shall allow user to create new customer and set his details.
- The system shall display the customer support contact numbers on the screen.
- The system shall allow user to enter the contact number for support personnel to call.
- The system shall display detailed invoice for invoice once it is confirmed.
- The system shall optionally allow user to print or save the invoice to specific customer.

1.5. Allow multiple payment methods.

- The system shall display available payment methods for payment.
- The system shall allow user to select the payment method for invoice.

1.6. Suppliers screen

- The system shall allow manager to add new supplier.
- The system shall allow manager to edit current supplier.
- The system shall allow manager to search for supplier.
- The system shall display available supplier to the casher on products screen and user casher can select supplier and add to product.

1.7. Sales Report

- The system shall allow manager to view sales report on specific days or date range include cash income and total profit.

SYSTEM USER REQUIREMENTS

Casher module:

1. can access POSWA system.
2. logging into the system.
3. view POS Screen.
4. appearing for the manager/admin.
5. can change the password.
7. resetting of forgotten password.
8. view/modify profile.

Administrator module

1. Back up for the database.
2. Full control to manage the system.
3. Create/delete/manage/update/view user account.
4. Editing/Deleting/Creating the records.
5. Add manager and casher.
6. Can change the password.
7. Log all event for user.
8. Record/manage transaction and records for all user.
9. system should support WAN so that POSWA can access from LAN.

Manger module

1. Logging into the system.
2. manage suppliers, products.
3. Add casher account.
4. Adding/editing/deleting new product.
5. Adding product to a category.
6. Adding/editing/deleting category.
7. Adding/editing/deleting suppliers.
8. View sales report.
9. change password.
10. View/modify profile.

System module

1. auto backup system.

2. Tax management enable/hide Tax with bill.

THE SYSTEM QUALITY REQUIREMENTS

1. Performance:
    - The POSWA shall be based on web and has to be run from a web server.
    - The POSWA shall take initial load time depending on internet connection strength which also depends on LAN or over internet that the system is run.
    - The performance shall depend upon hardware components of the client/customer.
    - After completing shopping process, the system will calculate the total amount to be paid and print it out along with the details on a receipt.
    - The system also provides an accurate inventory summary allowing the admin to easily track the stock contents and make an order if a product is to be out of stock.
    - At the end of the day, it provides the admin with a summary about the sales and the profits earned through the entire day.
    - Furthermore, the system supports multiple user at a time.
2. Usability:

Graphical User Interface

- The application should be user friendly and requires the minimal efforts to run and manage.
- The POSWA shall provide a uniform look and feel between all the system page.
- The POSWA shall provide use of icons and toolbars.

Accessibility

- The POSWA shall provide multi language support but in version 1 English language will support and Arabic language will add in version 2.

3. Portability:

The application is to be developed using PHP, CSS, HTML, JSP, and MySQL, which are platform independent technologies, making it easy to transfer and run on different platforms with minimum system requirements.

4. Availability:
   - The seller can add new products with all related details, and track the stock contents easily while working in the store.
   - For each purchase, the seller gathers and calculates the amount to be paid which is seen directly by the customer through a screen.
   - As it is a web application, the system can be accessed remotely by the admin to keep tracking the work progress.
5. Resolution:

In terms of accuracy, the system is characterized by its accurate outputs provided for sellers and admin.

6. Reliability:

The application is described to be as reliable as possible in terms of errors and lack of run failures, which may affect its operation, or cause it to hang out.

7. Integrity:

None of unauthorized persons can access this system, because the system prevents anyone to enter unless he/she was authorized in advance by the system admin.

8. Reusability:

The system can be reused in all stores and markets, and not limited to a certain ones.

9. Security:
   1. For security reasons, only the admin can manage and grant permissions to access different levels of system, and he is the only one responsible for password changing, while the staff can only access this system for sales, services and checking reports.
   2. The system shall automatically log out all user after a period of inactivity.

3. The system shall not use cookies with on the POSWA computer containing the user's or password.

POSWA INTERFACES

There are many types of interfaces as such supported by the POSWA software system namely; User Interface, Software Interface and Hardware Interface.

The protocol used shall be HTTP.

The Port number used will be 80.

1. User Interfaces

The user interface for the software shall be compatible to any browser such as Internet Explorer, Mozilla or Netscape Navigator by which user can access to the system.

The user interface shall be implemented using any tool or software package like php script, notepad++, adobe Dreamweaver etc.

2. Hardware Interfaces

Since the application must run over local area network or over the internet, all the hardware shall require to connect internet will be hardware interface for the system. As for e.g. network interface card, Modem, WAN, LAN, and Ethernet Cross-Cable.

3. Software Interfaces
    1. The POSWA system shall communicate with barcode scanner to rad the product code.
    2. The POSWA shall communicate with the product store to get the product specifications, offerings and promotions.
    3. The POSWA system shall communicate to credit management system for handling financing options.

DETAILED DESIGN CLASS DIAGRAM

POSWA software can summarize in simple steps for administrator, Seller and buyer as users will open the POSWA system, this system will install and configure on local computer or can be run from online server, after cashier login to the system he/she can open POSWA and POS screen and start to add products for customer after finish of shopping cart the cashier will add the amount of cash then print Receipt for customer, system user can add product for the system or update existing info and price for products, POSWA can view reports for date range include profit amount for selected date.

This POSWA is also aimed at specifying requirements of software to be developed but it can also be applied to assist in market or supermarket. The system can be used for small businesses they have a related user requirement's as POSWA had like most of modern and intelligent POS requirements needed by user in now days comparing this with many POS systems.

POSWA SYSTEM DESIGN

Depending in analysis documentation the system can view as five subsystems include cashier module, Administrator module, Manger module, and System module which include the two subsystems will save data and output to Database subsystem or local storage subsystem and all of subsystem interact with Manage System as view operating system Figure 6 describe the POSWA subsystem.

Figure 6. AEOS subsystem

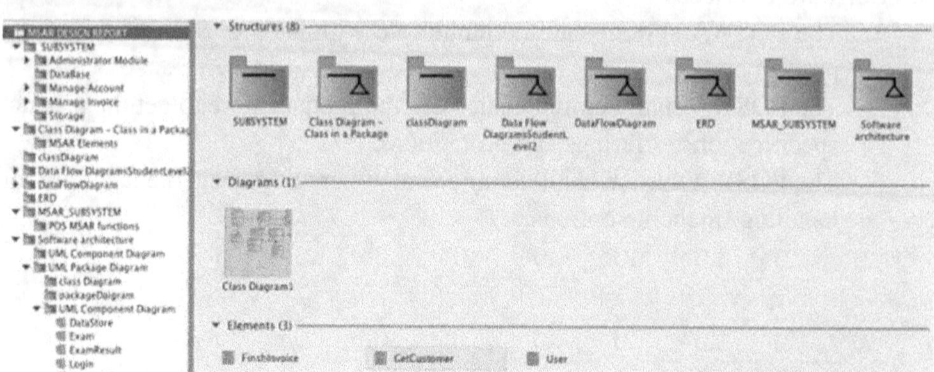

CLASS DIAGRAM

Class diagrams are a cornerstone of object-oriented software engineering, providing a static view of a system's structure. They visually represent the system's building blocks – the classes, their attributes, methods, and the relationships between them. By depicting these elements using UML notation, class diagrams serve as a blueprint for understanding the system's design, guiding developers in translating abstract concepts into concrete code.

Class diagram provides a comprehensive visual representation of the POSWA package structure. It employs UML notation to detail the classes, their attributes, methods, and the relationships that bind them together within the system. Figure 7, a class diagram for the POSWA package, exemplifies this principle. It meticulously outlines the classes involved in a point-of-sale web application, detailing their attributes and functionalities. The diagram also reveals the intricate relationships between these classes, such as association, inheritance, aggregation, and composition, clarifying how they collaborate to form a cohesive system. This visualization of the system's static structure is crucial for communication among developers, ensuring everyone understands the system's design and how different components interact, ultimately leading to a more efficient and maintainable software solution.

Figure 7 outline features several crucial classes, each representing a distinct element of the system. "Manager" and "Cashier" depict user roles with specific attributes and methods. "AddCustomer" encapsulates the process of adding customer information, while "UserLogIn" handles the system's user authentication. The "Admin" class represents the system administrator with overarching control. "ProductStore" manages the product inventory, including details about suppliers through the "Supplier" class. "Invoice" represents the sales transactions, and "Sales" likely reflects the system's sales records.

The relationships between these classes are crucial for understanding the system's functionality. "Association" represents a general connection, while "Inheritance" demonstrates a hierarchical "is-a" relationship. "Aggregation," a "has-a" relationship, is illustrated by "ProductStore" containing "Supplier." "Composition," a stronger form of "has-a," is shown with "ProductStore" composing "Sales," implying a dependent relationship. The diagram effectively maps out the interactions between different user roles, system functionalities like login, product management, and sales recording, and crucial data elements like customer, supplier, invoice, and sales information. This detailed representation serves as a valuable blueprint for comprehending and implementing the POSWA package, guiding developers and stakeholders in understanding its structure and functionalities.

Figure 7. Class diagram for POSWA package

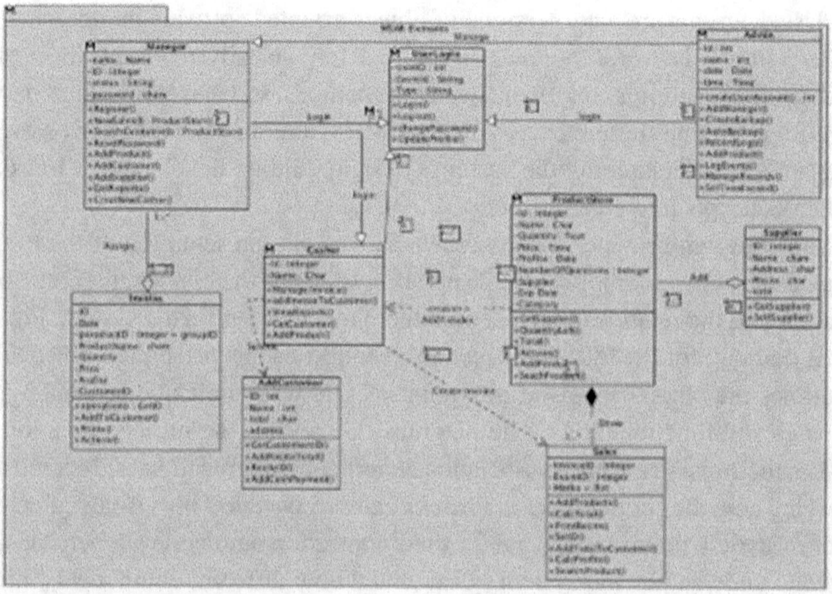

use case diagram illustrates the main functions and interactions of users within a system, which include three actors are identified: "Cashier," "Manager," and "Admin". Use cases Administrator:

1.Back up for the database.

2. Full control to manage the system.
3. Create/delete/manage/update/view user account.
4. Editing/Deleting/Creating the records.
5. Add manager and cashier.
6. Can change the password.
7. Log all event for user.
8. Record/manage transaction and records for all user.

Use cases cashier:

1. can access POSWA system.
2. logging into the system.
3. view POS Screen.
4. appearing for the manager/admin.
5. can change the password.

7. resetting of forgotten password.
8. view/modify profile.

Use cases Manger:

1. Logging into the system.
2. manage suppliers, products.
3. Add cashier account.
4. Adding/editing/deleting new product.
5. Adding product to a category.
6. Adding/editing/deleting category.
7. Adding/editing/deleting suppliers.
8. View sales report.
9. change password.
10. View/modify profile.

The "Cashier" interacts with functions related to day-to-day operations: "Login/Register," "Start New Sales," "change/reset password," "View/Edit product," "View Supplier," "Add Cashier," and "View Reports." The "Manager" has broader responsibilities, including all cashier functions plus "manage Products," "manage Store," "add Store," "Manage Records," "Backup DB," "logs all user events," and "add Manager." The "Admin" appears to have the highest level of access, managing users, the system, the question bank (which seems out of place in this context), the store, and products.

Figure 8 uses dashed lines to indicate optional interactions ("can") and solid lines for direct interactions. It also uses labels like "assign," "Export," "type," "perform," "run," and "save" to clarify the nature of the interactions between actors and functions.

Figure 8. Use case for main actors in system

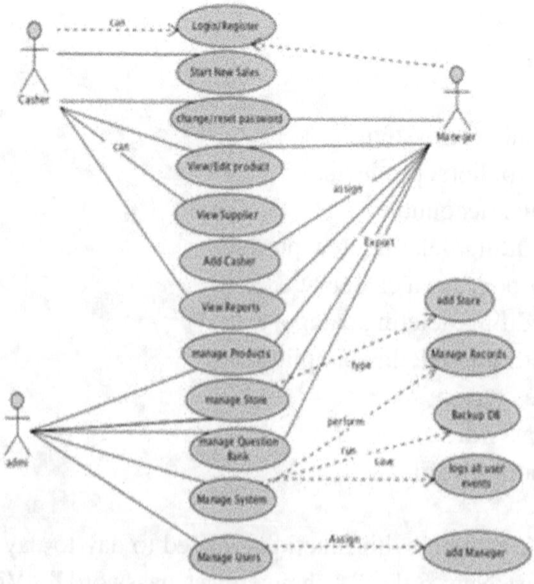

The "Manager" has access to functionalities related to managing products and suppliers. "ManageProducts," "Create Supplier," "View Reports," and "Manage User" are core use cases. "ManageProducts" is further extended by "Supplier" and "NewProduct" use cases, while "Create Supplier" is extended by "Products" and "Set SupplierTotal". The "Cashier" interacts with the system primarily for sales-related tasks. The "Start Sales" use case represents the initiation of a sales transaction, while "Check Out" signifies its completion. Both of these use cases include functionalities from "Login Account" and "Products," suggesting that the cashier needs to be logged in and access product information to perform sales transactions.

Figure 9 illustrating the interactions between a "Manager" and a "Cashier" in a system for managing products and suppliers. The diagram utilizes UML notation to depict the actors, use cases, and their relationships. effectively highlights the distinct roles of the "Manager" and the "Cashier" and the use cases associated with each. The use of "extend" relationships clarifies how certain use cases add functionalities to others, providing a detailed view of the interactions and dependencies between different functions within the system.

Figure 9. Manager and cashier use case

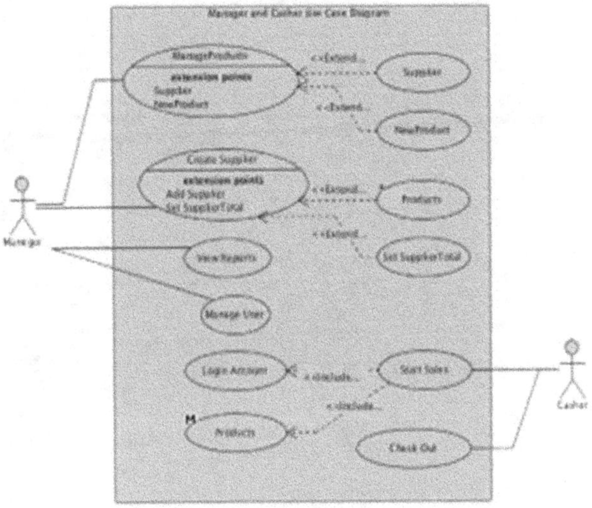

Administrator use case diagram highlights the functionalities and interactions of the "admin" within a system, it showcases the various actions the administrator can perform, demonstrating their high level of access and control. The administrator has access to a wide range of functionalities:

- Manage Users: Overseeing user accounts, including creation, modification, and deletion.
- Manage System: Controlling system-wide settings and configurations.
- Manage Question Bank: Managing the repository of questions, which seems out of place in this context and might be a remnant from a different system.
- Manage Store: Overseeing store-related details, inventory, and operations.
- Manage Products: Managing products, including adding, editing, and potentially removing products.
- Manage Records: Managing system records, potentially including sales, inventory, or user activity logs.
- Backup DB: Creating backups of the database to ensure data security and recoverability.
- Logs all user events: Monitoring and logging all user activities within the system for auditing and analysis.
- Add Manager: Creating new manager accounts and assigning roles

Figure 10. Administrator UML case diagram

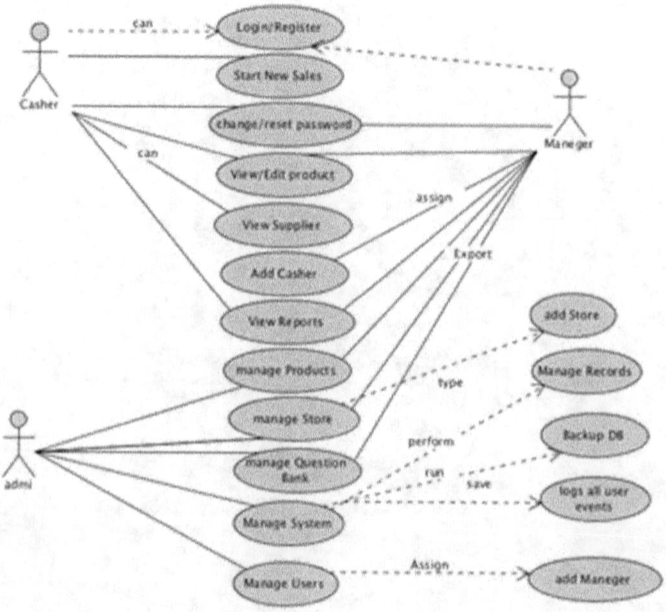

Figure 10 shows the administrator's potential interactions with functionalities typically associated with other roles: "Cashier" and "Manager." This suggests that the administrator has the authority to perform actions like "Login/Register," "Start New Sales," "change/reset password," "View/Edit product," "View Supplier," "Add Cashier," "View Reports," "add Store," and "Export." The use of dashed lines for these interactions indicates they are optional, while solid lines represent direct interactions. Labels like "assign," "perform," "run," "save," and "type" further clarify the nature of the interactions. Table 1 effectively depicts the administrator's overarching role in the system in details for each use case describe in Figure 10, showcasing their ability to manage users, the system, and various data elements. It underscores the administrator's high level of access and control, allowing them to perform a wide range of critical tasks.

Table 1. Case diagram details

Use Cases Code:	**MASR01**
Use Cases Name:	Create new account

continued on following page

Table 1. Continued

Use Cases Code:	**MASR01**
Casual Description:	
Use Case: Log-in The user must be on the log-in form. The user must enter his/her authorize username and password.	
Use Cases Code:	MASR02
Use Cases Name:	Update User
Casual Description:	
Use Case: Update User The user must logged-in as administrator. The user must search and double-click the user's information under the data grid to be able to retrieve and update his/her data.	
Use Cases Code:	MASR03
Use Cases Name:	Manage Product
Casual Description:	
Use Case: Manage Product The user must be logged in as administrator. The user must select the Manage Product option from the Data Management window.	
Use Cases Code:	MASR04
Use Cases Name:	Add Product
Casual Description:	
Use Case: Add Product The user must be logged in as administrator. The user must select the Add Product option from the Manage Product window.	
Use Cases Code:	MASR05
Use Cases Name:	Edit Product
Casual Description:	
Use Case: Edit Product The user must be logged in as administrator. The user must search and double-click product's details under the data grid to be able to retrieve and edit the product.	
Use Cases Code:	MASR06
Use Cases Name:	Delete Product

Casual Description:
Use Case: Delete Product
The user must be logged in as administrator.
The user must search and double-click product's details under the data grid to be able to retrieve and delete the product.

Cashier use case diagram outlining the functionalities and interactions of a "Cashier" within POSWA system, it uses a blue rectangle to encapsulate the system's context, with oval shapes representing use cases and a stick figure representing the

"Cashier" actor. The "Cashier" interacts directly with three main use cases: "System," "Product Managment," and "Report Manegment."

- System: This use case likely represents basic system-level actions. It includes "login" and "Logout" functionalities, enabling the cashier to access and exit the system securely.
- Product Managment: This use case covers functionalities related to managing products. It includes actions like "add product to sales" and "sales Del product," enabling the cashier to add or remove products from a sales transaction. It also has an association with "ProductSearch," suggesting the cashier can search for products within the system.
- Report Manegment: This use case focuses on report generation. It includes actions like "add" a new report and generating a "sales Report" with a "date."

The dashed lines in the Figure 11 indicate relationships between use cases, likely "extend" or "include" relationships. For instance, "add product to sales" extends from "Product Managment," suggesting it's a sub-functionality within product management. Similarly, "sales Report" extends from "Report Manegment," signifying a specific report type within report management.

Figure 11. Cashier UML case diagram

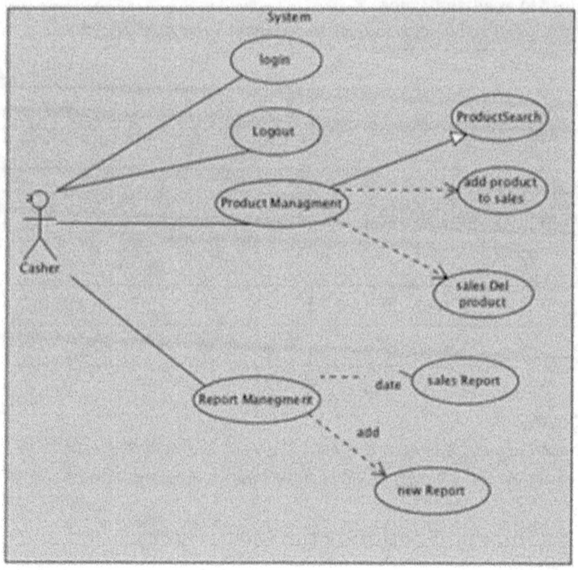

Table 2 effectively highlights the cashier's primary interactions with the system in details, encompassing system access ("System"), product management ("Product Management"), and report generation ("Report Management"). It provides a clear and concise visual representation of the cashier's role and functionalities within the point-of-sale system.

Table 2. Cashier use case details

Use Cases Code:	**MASR07**
Use Cases Name:	View Reports
Casual Description:	
Use Case: View Reports 1. The user must be logged in as cashier or administrator. 2. The user must select the View Reports option from the Data Management window.	
Use Cases Code:	MASR08
Use Cases Name:	View Sales report
Casual Description:	
Use Case: View Sales Transaction 1. The user must be logged in as cashier or administrator. 2. The user must select the Sales Transaction option from the View Reports window.	
Use Cases Code:	MASR09
Use Cases Name:	Add Product on Purchase List
Casual Description:	
Use Case: Add Product on Purchase List 1. The user must be logged in as cashier. 2. The user must select the product name under the product list from the Point of Sale window.	
Use Cases Code:	MASR010
Use Cases Name:	Delete Listed Product
Casual Description:	
Use Case: Delete Listed Product 1. The user must be logged in as cashier. 2. The user must select the product under the product list and select Delete Product option from the Point of Sale window.	
Use Cases Code:	MASR011
Use Cases Name:	Check Out

Casual Description:
Use Case: Check Out
1. The user must be logged in as cashier.
2. The user must select the Check Out option from the Point of Sale window.

SYSTEM ARCHITECTURE AND DYNAMIC MODELS

POSWA include set of components interact with actors this section describe system into components and how we Divide to MASR POS system, next we try to Explain the functions of each component and the relationships between the components.

Sequence Diagram

Sequence diagram for user login, illustrates the process of user login, outlining the steps and interactions between the "User," the "Login page," the "System DB (Users)," and the "user page display." The user initiates the process by entering their username and password on the login page. This information is then passed to the system database for validation. If the credentials are valid, the database confirms this validation, and the system then displays the user's designated page, indicating successful login. Figure 12 effectively showcases the sequential flow of actions and information exchange during the user login process, highlighting the key role of the system database in verifying user credentials and granting access.

Figure 12. Sequence diagram for user login

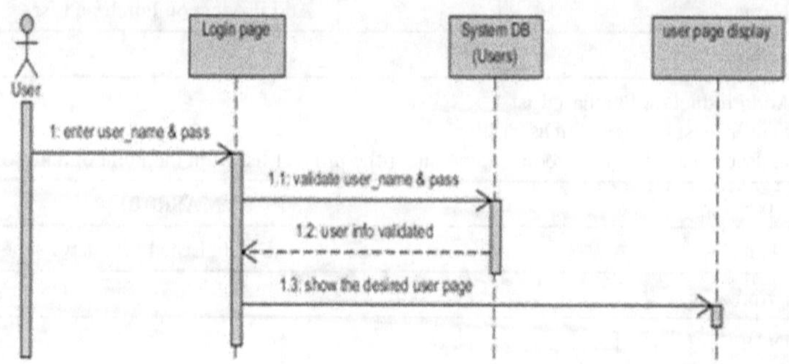

Add new manager an activity diagram, illustrates the process of adding a new manager to POSWA system. Figure 13 depicts the interactions between the "Admin," the "add new manager page," and the "system DB(users)." The process begins with the admin entering the new manager's information on the designated page. This information is then sent to the system database, which stores the new manager information. Upon successful storage, the system confirms that the new manager information has been added, concluding the process. This diagram effectively captures the sequential flow of actions involved in adding a new manager, highlighting the key role of the system database in storing user information.

Figure 13. Add new manager activity diagram

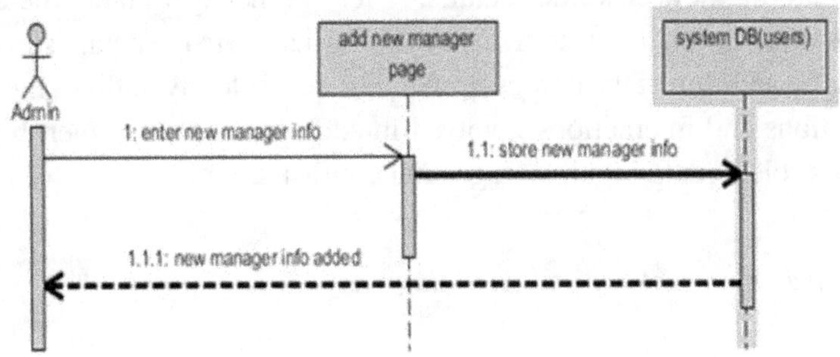

Add new cashier activity diagram outlines the steps involved in adding a new cashier to POSWA, the "manager" acts as the initiator, entering the new cashier's information on the designated "add new cacher page." This information is then sent to the "system DB," representing the system's database, where the new cashier information is saved. Upon successful storage, the system confirms that the new cashier has been added, concluding the process. Figure 14 effectively captures the sequential flow of actions and the interaction between the manager, the system interface, and the database.

Figure 14. Add new cashier activity diagram

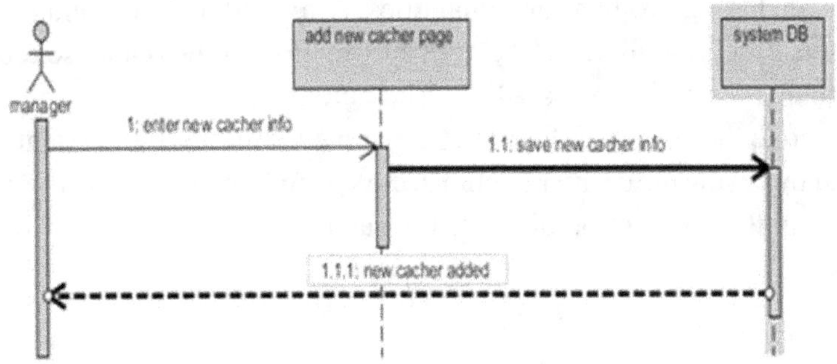

Add new customer Activity Diagram visualizes the step-by step process of adding a new customer within POSWA, it depicts the actions taken by the "manager," the system's interface elements ("customer page," "add customer button," and "save button"), and the "system DB" (database). The process begins with the manager initiating the addition of a new customer. The system then provides the necessary fields for entering new customer details. After the manager enters the details and presses the save button, the system saves the information to the database, confirming the successful addition of the new customer. Figure 15 clearly outlines the sequential flow of actions and interactions involved in adding a new customer, highlighting the system's role in data input, storage, and confirmation.

Figure 15. Add new customer activity diagram

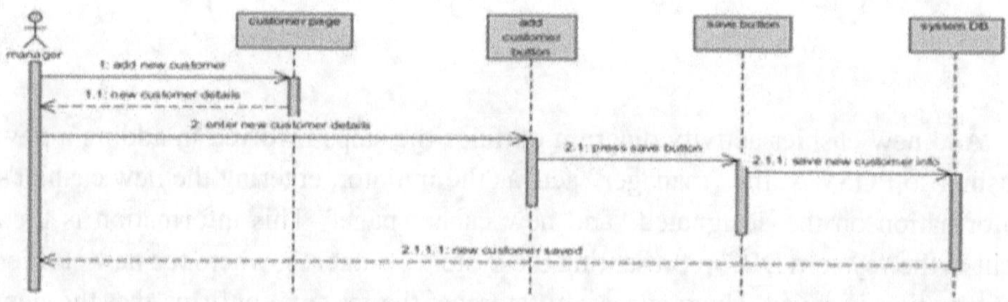

Add new sale activity illustrates the process of adding a new sale in POSWA. The diagram highlights the actions of the "casher," the system interface ("sales page," "Add button," "Save button," "print receipt button"), the "system DB" (database), and the "customer." The process begins with the casher selecting the new sale parameters, including products and quantities. After adding items and pressing the "Save button," the system saves the sale information to the database, confirming the sale. Finally, the casher presses the "print button" to generate a receipt, which is then delivered to the customer. Figure 16 effectively captures the sequential flow of activities, showcasing the cashier's actions, the system's response, and the customer's involvement in the completion of a sale transaction.

Figure 16. Add new sale activity diagram

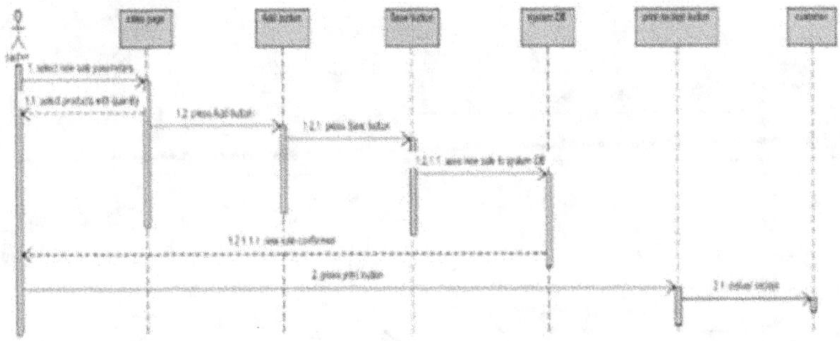

Add new product activity diagram illustrates the process of managing products within POSWA, it highlights the actions of the "manager" and the system's response, represented by the "products page" and "system DB" (database). The manager can perform three main actions: add a new product, edit an existing product, or delete an existing product. For each action, the diagram outlines the sequence of steps: the manager provides input, the system responds by requesting details or saving information to the database, and the system confirms the action's completion. Figure 17 effectively captures the flow of events and the interaction between the manager and the system in managing product data.

Figure 17. Add new product activity diagram

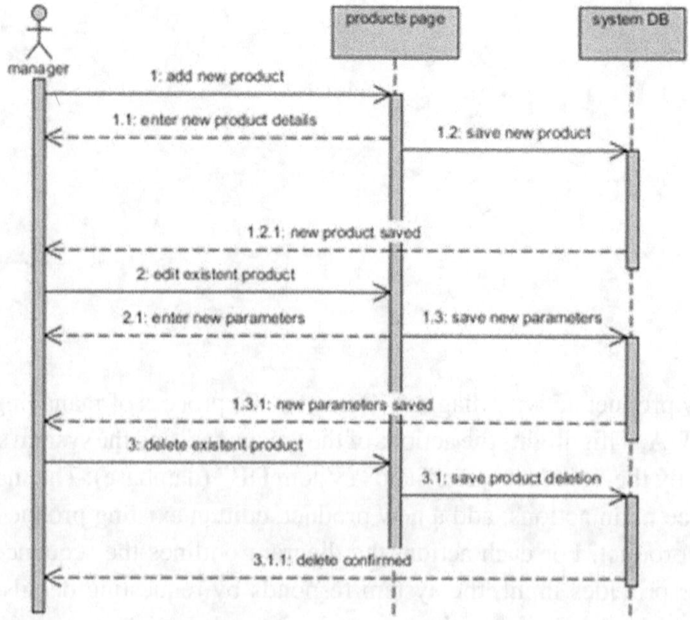

Add new supplier activity illustrates the process of adding a new supplier to POSWA, it showcases the sequential steps involved, highlighting the interactions between the "manager," the system interface ("suppliers page," "add supplier button," "save button"), and the "system DB" (database). The manager initiates the process by indicating they want to add a new supplier. The system responds by displaying the "suppliers page" and prompting for new supplier details. Once the manager enters the details and clicks the "save button," the system saves the new supplier information to the database, confirming the successful addition. Figure 18 effectively captures the flow of actions and data exchange between the manager and the system, providing a clear visual representation of the "Add New Supplier" process.

Figure 18. Add new supplier activity diagram

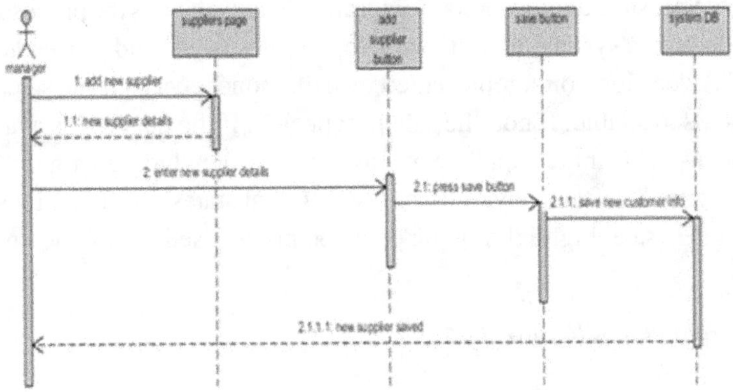

Issue sales report activity diagram illustrates the process of issuing a sales report within a system POSWA it outlines the steps involved, highlighting the interactions between the "manager," the system interface ("sales report page," "search button," "print button"), and the "system DB" (database). The process begins with the manager selecting the "sales report page" and specifying a time interval for the report. The manager then presses the "search button," prompting the system to retrieve the relevant data from the database. Once the data is retrieved, the manager presses the "print button" to generate a printed copy of the sales report. Figure 19 effectively depicts the sequential flow of actions and data retrieval, showcasing the manager's role in initiating the report generation process and the system's role in retrieving and presenting the data.

Figure 19. Issue sales report activity diagram

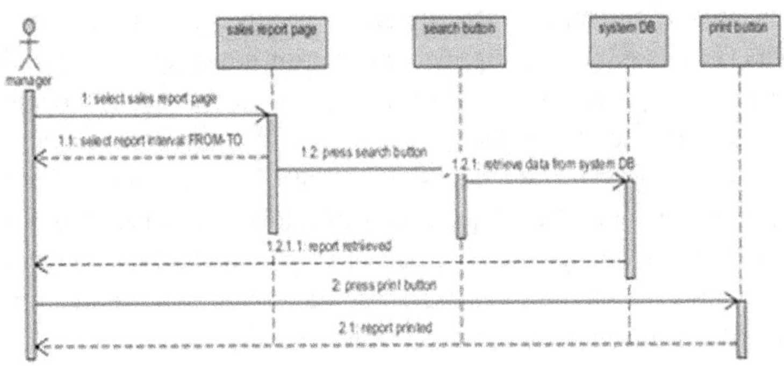

State Diagram

Add new Sale state diagram a state diagram, visually depicts the process of adding a new sale within a system. It starts with "open sales page" and proceeds to "select products." A decision point, represented by a diamond, checks if the selected product quantity is available. If not, the sale is canceled. If the quantity is available, the process moves to "Print receipt," then "Save to DB," concluding with the "End sale activity". Figure 20 effectively outlines the different states and transitions involved in adding a new sale, highlighting the decision point based on product availability.

Figure 20. Add new sale state diagram

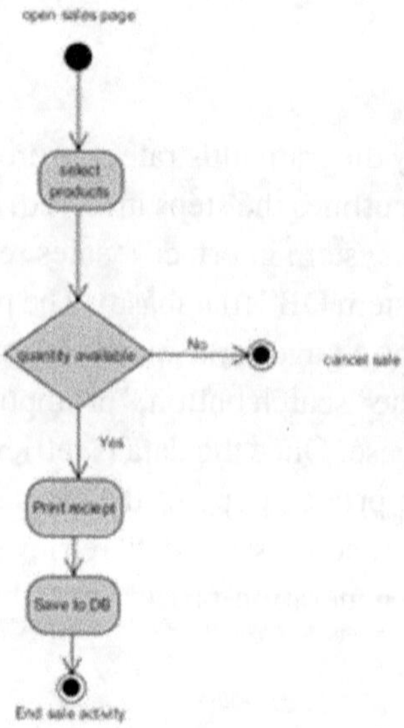

Add new product and supplier state diagram visually depicts the process of adding a new product and its associated supplier information in a system. The process begins with "open products page," followed by "Add new product details." A decision point, represented by a diamond, checks if the supplier for the product already exists. If not, the user is prompted to "add new supplier info." If the supplier exists, the user "confirm new product info." Both paths converge at a point where the system "save new data to DB," concluding with "End activity". Figure 21 effectively illustrates the different states and transitions involved in adding product and supplier information, highlighting the decision point based on supplier existence.

Figure 21. Add new product and supplier state diagram

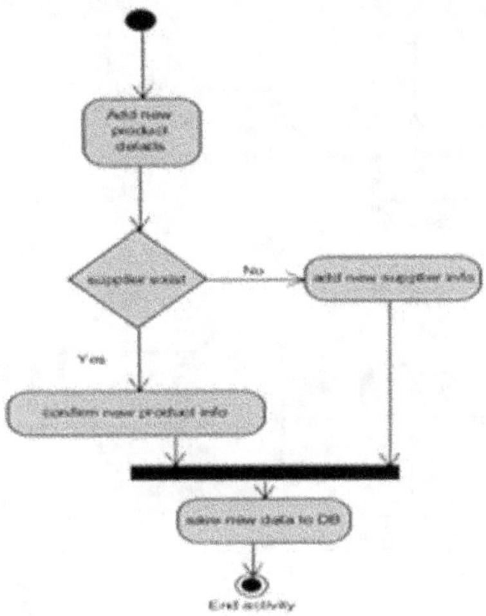

State Diagram for POSWA Users main navigation illustrating the user's main navigation path within an AOES. Starting from “Open System,” users can either register (“Register”) or fill in their UserID and password to “Login.” After a successful login, the system branches based on the user's selected role – “Cashier,” “Admin,” or “Maneger” (likely “Manager”). Cashiers can “open Sales,” “Start add Products,” and either “print Invoice” or proceed to “chekout.” Admins can “Add User” and then “Logout.” Managers can “Manage System” and “Add Products,” eventually leading to “Logout”. Figure 22 effectively showcases the different paths and states available to users based on their roles and actions within the system.

Figure 22. User main navigation path for the AEOS system

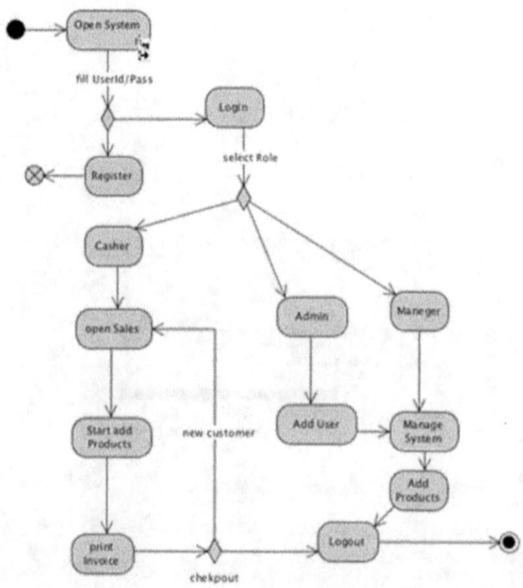

Software Architecture

Software architecture presents a comprehensive overview for a system. The diagram showcases various components, including use case diagrams, activity diagrams, data flow diagrams, state diagrams, and class diagrams, organized hierarchically within folders. It highlights the interconnectedness of different modeling techniques used to capture various aspects of the system, from high-level use cases to detailed data flows and state transitions.

The top section focuses on design reports, listing various diagrams and their hierarchical relationships. For example, the "MSAR_SUBSYSTEM" folder includes an activity diagram, a class diagram, and a use case diagram, suggesting a progression from high-level use cases to more detailed design elements.

The bottom section provides a visual representation of the system's architecture, linking specific functionalities to different user roles (admin, cashier, manager, user) and system components. For instance, "Add Products" is linked to the "Admin" role, while "Start New Sales" is associated with the "Cashier" role. This section effectively maps out the functionalities of the system and how they are distributed across different user roles and system elements.

Overall, Figure 23 offers a valuable bird's-eye view of the POSWA software architecture, showcasing the different modeling techniques used to capture the system's structure, behavior, and data flow. It underscores the importance of a systematic approach to software development, using a variety of diagrams to represent the system from different perspectives and levels of detail.

Figure 23. POSWA software architecture

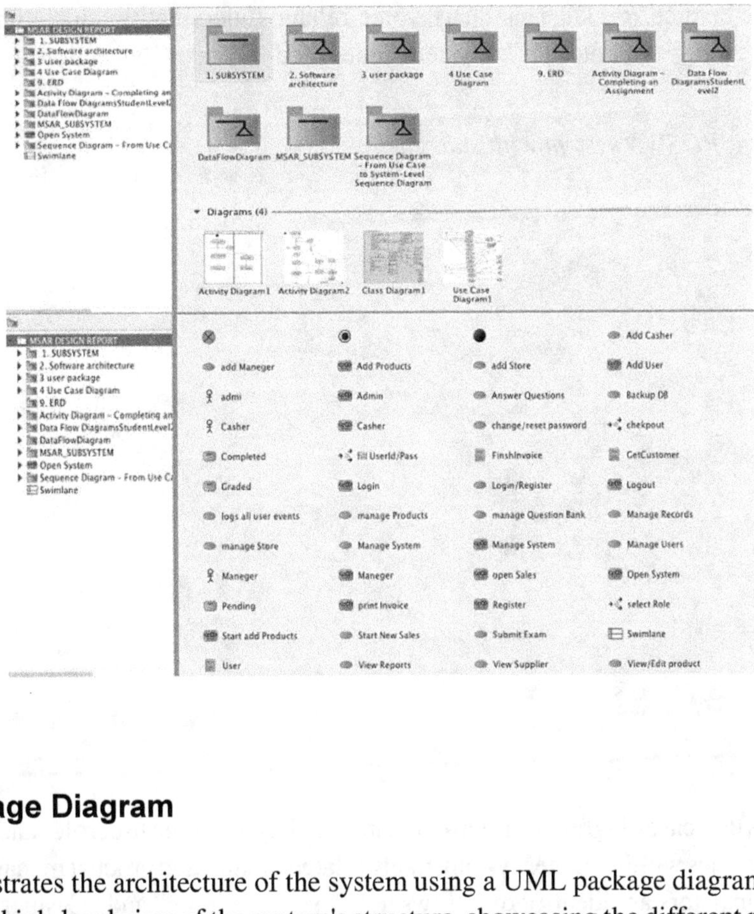

Package Diagram

illustrates the architecture of the system using a UML package diagram, it provides a high-level view of the system's structure, showcasing the different packages and their dependencies. The system is organized into several interconnected packages, each representing a distinct area of functionality. package diagram effectively visualizes the system's structure, highlighting its modularity and the relationships between different functionalities. It provides a clear and concise overview of the POSWA system's architecture, promoting understanding and facilitating development and maintenance, it also illustrates dependencies between these packages. "Update-Profile" connects the "Administrator Module" to the "Manage Account" package, while "CreateInvoice" links the "Administrator Module" to the "Manage Invoice" package. Additionally, "save Data" connects the user roles to both "Storage" and "Database," suggesting data persistence mechanisms.

Figure 24 contains sub-packages for "Administrator Module," "Manage Account," and "Manage Invoice." The "Administrator Module" houses the core user roles: "ADMINISTRATOR," "Cashier," "USER," and "Maneger," indicating different levels of access and privileges within the system. The "Manage Account" package focuses on user account management, with a "User Account" element, while the "Manage Invoice" package handles invoice-related functionalities, including "Customer," "Add_paymentCash," "SystemTime," and "SystemConnection."

Figure 24. POSWA system with subsystem

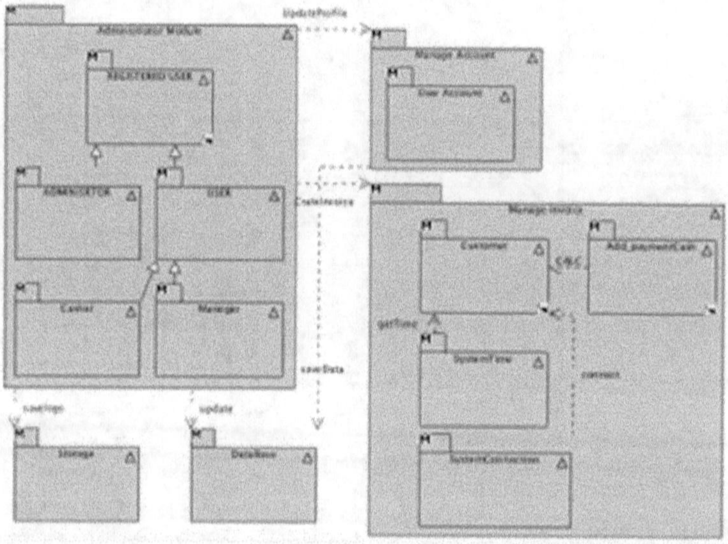

Users use-case diagram showcases the interactions between user roles and system functionalities within a larger system, likely related to sales and product management. Four main actors are identified: "Browser," "User," "Admin," and "Cashier." The " Browser " role appears unconnected to the main functionalities, suggesting it might be a remnant from a previous diagram or an unrelated entity. The central "User" role branches into "Admin," "Cashier," and "Manager," highlighting the different user types within the system.

Users use case divided into two packages: "User Package" and "System Package." The "User Package" focuses on user management tasks, with use cases like "Manage User," "EditUserDetails," "add manager," and "add casher." The "System

Package" encompasses broader system functionalities, including data management ("Backup," "createBackup," "UpdateProfile," "AutoBackup"), sales and product management ("Sales," "manage suppliers," "Manage Products"), event logging ("Log all event"), general management ("Manage," "Record"), and time management ("Time management").

The diagram effectively maps out the interactions between user roles and system functionalities, illustrating the different actions each user type can perform. The "Admin," "Cashier," and "Maneger" roles interact with various use cases within the "System Package," highlighting their specific responsibilities and privileges within the system, Figure 25 provides a clear visual representation of the system's functionality and the roles of different users within it, as shown solid lines depict direct interactions between actors and use cases, while dashed lines likely represent "extend" or "include" relationships, showing how certain use cases add functionality to others or are included within specific functionalities. The "Admin," "Cashier," and "Manager" roles interact with a range of use cases within the "System Package," reflecting their specific responsibilities and access levels. This visual representation of user interactions with system functionalities provides a clear and concise overview of the system's features and user roles.

Figure 25. Users case package interact

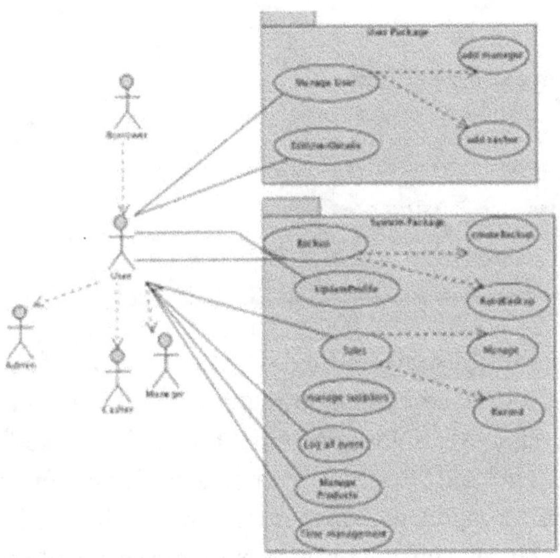

The Administrator has access to twelve distinct elements: "add product," "add product info," "Delete Product," "edit product," "login," "Logout," "new Report," "Product Managment," "product report," "ProductSearch," "Report Manegment,"

and "sales Report." These elements suggest the administrator has full control over product management, report generation, and user authentication. Cashier has access to ten elements: "add product," "add product to sales," "edit product," "login," "Logout," "new Report," "Product Managment," "ProductSearch," "Report Manegment," and "sales Report." These elements indicate the cashier's role in managing sales, adding products to sales transactions, searching for products, and generating reports. Notably, the cashier also has access to "add product," "edit product," and "Product Managment," suggesting a level of product management responsibility beyond basic sales transactions.

This breakdown of functionalities for each role provides valuable insights into the system's design, highlighting the different levels of access and control granted to each user type.

Figure 26 showcases the main package listed above for both the Administrator and Cashier roles within a system.

Figure 26. Administrator and cashier sub system

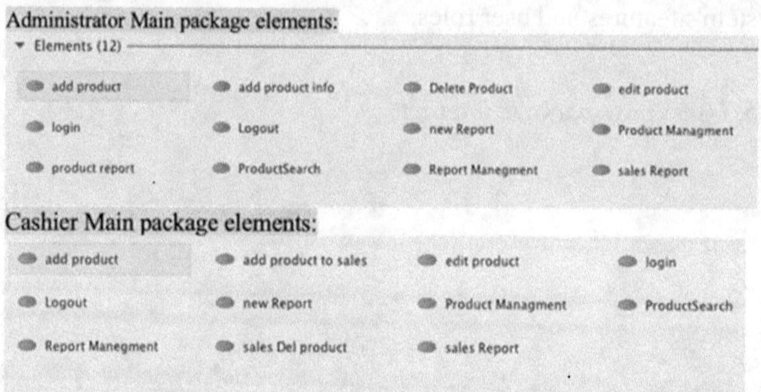

Component Diagram illustrates the high-level architecture of a system, by depicting its key components, interfaces, and their interactions. The "DataStore" component likely represents a database, holding persistent data. The "User" interacts with the system through "Login" and "Register" components for authentication and account creation. The "Sales" component manages sales transactions, utilizing the "GetCustomer" interface to retrieve customer information from the "Products" component and the "FinishInvoice" interface to trigger the "PrintInvoice" component for generating invoices. This diagram effectively showcases the system's modular design, where components interact through well-defined interfaces, promoting flexibility and maintainability. Figure 27 outline relations for component diagram using UML.

Figure 27. UML component diagram

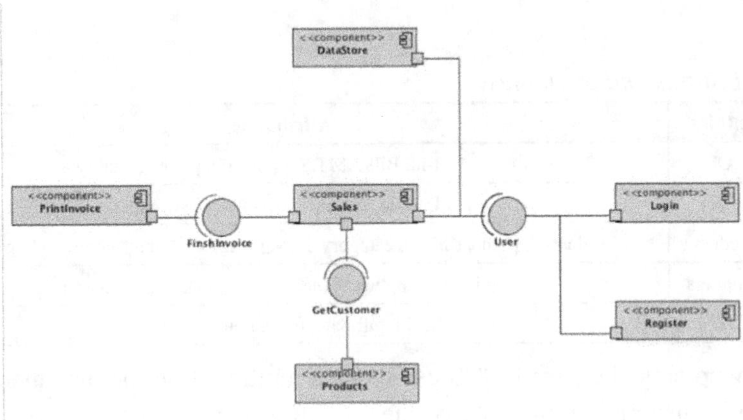

Entity Relationship Diagram

An Entity-Relationship (E-R) Diagram helps us to explain, diagrammatically, the overall high-level structure of a database as concluded by Silberschatz (1997). Elmasri & Navathe (1999) state an E-R diagram can be used as a reference to ensure all users' requirements are met and that no conflicting requirements occur.

Figure 28. The conceptual model of the system database

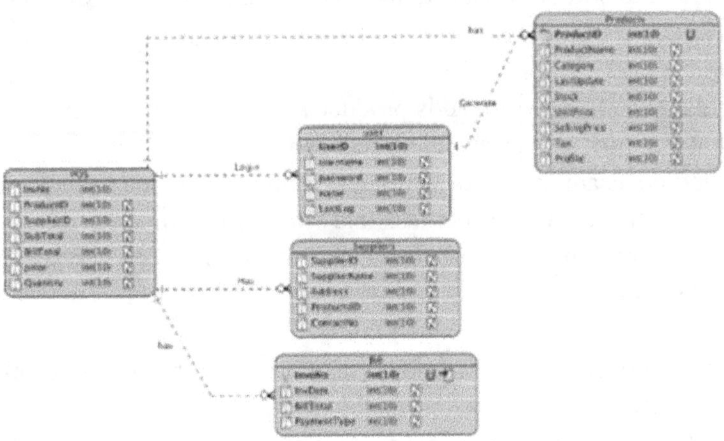

Figure 28 shows the relationship between every entities and attributes. which we have many entities and each entity it has many attributes as listed in Table 3.

Table 3. Entities and attributes

NO	Entities	Attributes
1	POS	Product-id, Bill-AMT,Supplier-ID, price, Quantity
2	user	User-id, Name, username, password,
3	Product	Product-id, product-name, category, last-update, stock, unit-price, sealing price
4	Suppliers	Supplier-id, supplier-Name, Address, material, contact-no
5	Bill	Bill-id, bill-date, bill-amount, payment

For this practical example POS can receive multiple User or customers, which consist user/customer or more, Also, one user/customer or more can be login to the system. POS has multiple products as consist of many products and consist of many bills, which there are many products in POS system, POS has many suppliers, that's mean there are one or more suppliers, in other hand many suppliers have multiple products, and many suppliers maybe have one or more products, finally A user or customer has one or multiple bills.

IMPLEMENTATION AND TESTING

In this section we will describe the sequence that the user should be followed to use, practical example we use in this chapter have include three main users in the system (administrator, Manager, and cashier). Each user has different roles, as shown follow:

1. Administrator: Super user, adds products, add supplier, and manages system, do all admin task and access all system feature.
2. Manager: do as administrator without system manage and cannot add user.
3. Cashier: User mostly a cashier task and with limited function to system.
4. System: the system does automatic task that assign by user and should execute depend on time or events.

After the user open using internet browser (Chrome, Firefox, Microsoft Exp.) the system main page display to user as main Login page, then depending on the user role user will login to the system and display user main menu which different and separate to three groups Administrator, Manager and cashier as shown in Figure 29.

Figure 29. Login main page

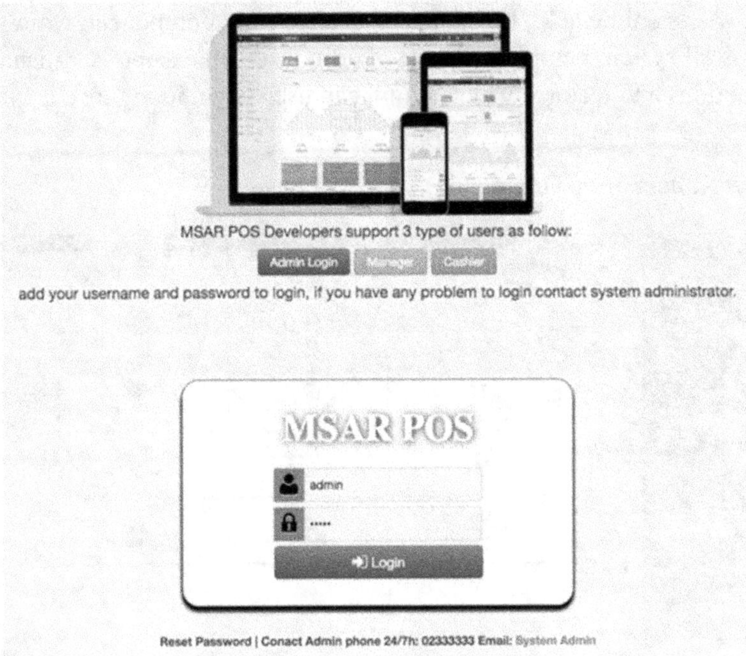

From main page, users can login to Main POS panel by filling in the Username and Password in main login page, and If user add error credential, error massage will be displayed to user as in Figure 30, and if user use blank data in form, then error massage will display to user.

After user login to POSWA and system validate correct credential info username and password, then info pass using session variables to main page, using simple algorithm it generates automatic page for each user type as we described previously.

Figure 30. Error massage for user account

Cashier main page generates after a user login successfully to the system, cashier can access system using username and password and depending on user category can view sales, online help, logout and exchange rates component, while manager can access all system component most of admin screen except user manage page. Main functions and features which admin can use shown in Figure 31.

Figure 31. Administrator main page

Sales screen available to Administrator, Manager and Cashier, allowing user to do most of daily works and this is the most important screen in our system as it used for sales and customer. Users can search for products by using the barcode reader or search by name, code and ID of the product as shown in Figure 32.
Main Functions for Sales Page are:

- Search product
- Add products to sales page include main product info with profit amount.
- User can add or drop products from sales table.
- Print bills and add to customer by name.
- Save invoice logs table.

Figure 32. Sales page

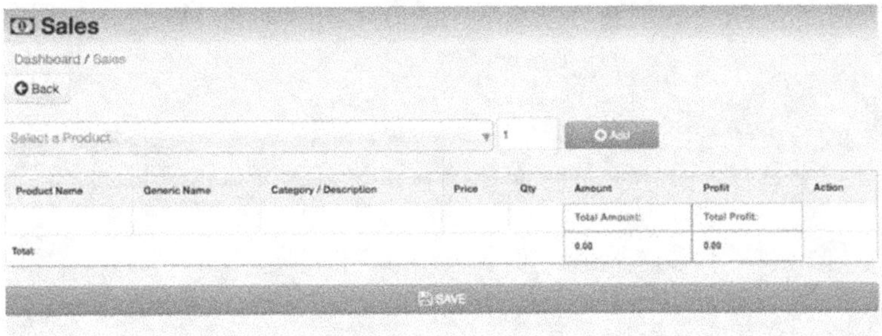

Test used in this page if user add quantity without name, the product will not be added to invoice, unless user will select it from product list as shown in Figure 33.

Figure 33. Error massage for add product not named in list

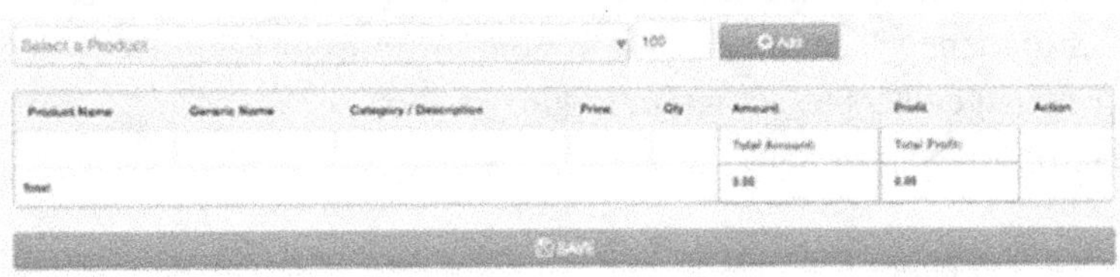

Product screen available for Administrator and Manager only, allowing them to perform most of daily work smoothly and this is one of the mostly visited screen in our system as it used for all products and their prices that are available in the system as shown in Figure 32.

Main Function for Sales Page:

- Search product
- Show all product in product page.
- Add new product with all info related to product.
- Edit product.
- Delete product.

Figure 34. Product page

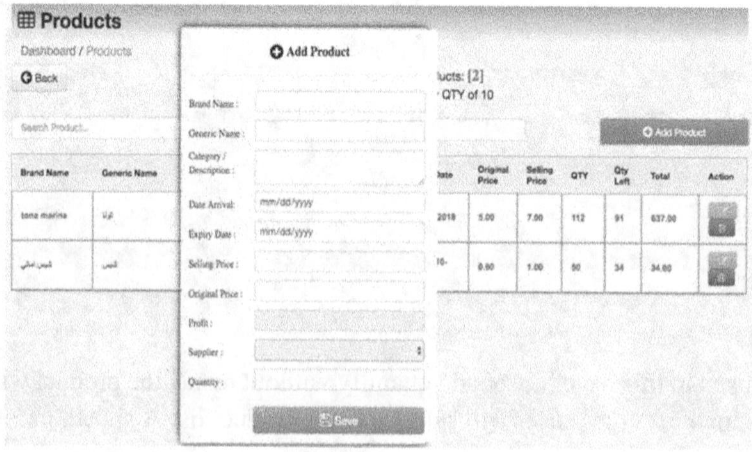

User screen available to administrator only, administrator can add user and assign roles to user or manage accounts on the system. Admin can search for a desired user account or manage users as shown in Figure 35.
Main Function for Sales Page:

- Search user
- Show all user in user page.
- Add new user with all info related to user.
- Edit user.
- Delete user.

Figure 35. User page

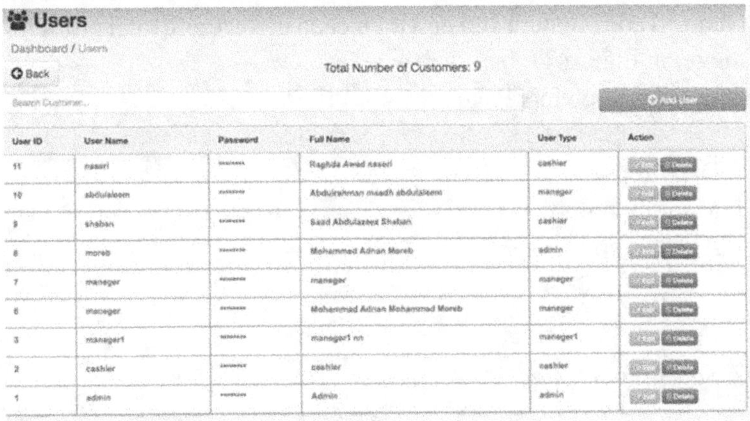

Suppliers screen available to administrator and manager only, they can add new Supplier and view or manage current Supplier as shown in Figure 35.
Main Function for Suppliers Page:

- Search for Supplier.
- Show all Supplier sin Supplier page.
- Add new Supplier with all info related to Supplier.
- Edit Supplier.
- Delete Supplier.

Figure 36. Supplier page

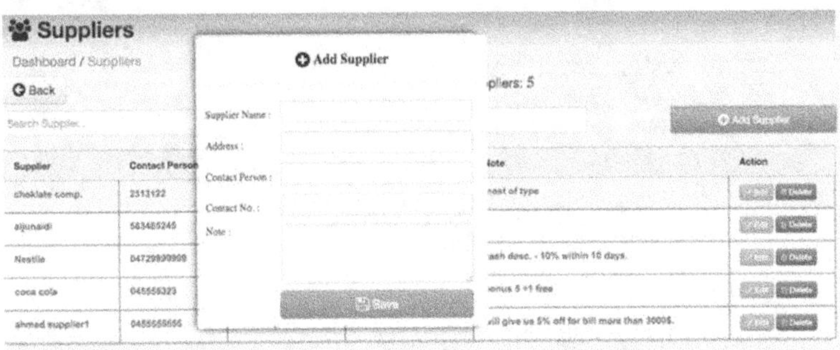

Report screen available to administrator and Manager only, they can view daily reports or report for date from to selected date the report include invoice number, customer name if add, amount and profit for each invoice and total profit for selected date, as shown in Figure 37.
Main Function for Suppliers Page:

- Daily report.
- Monthly report.
- Annual report

Figure 37. Sales report page

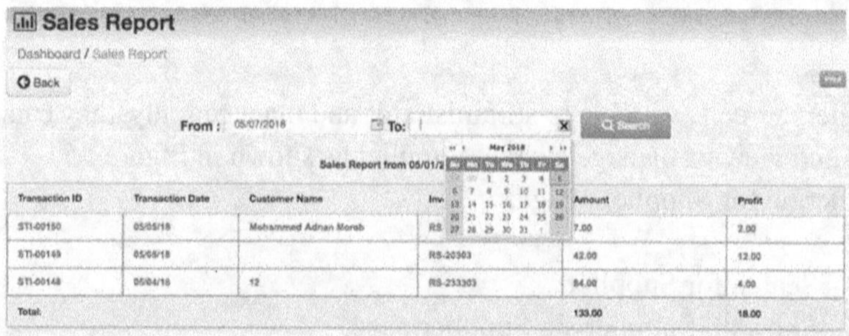

Product inventory report screen available to administrator and Manager only, they can view daily logs foe all events on the system, as shown in Figure 38.
Main Function for Suppliers Page:

- Search invoice-by-invoice number.
- Delete report.
- View all system events and transactions.

Figure 38. Product inventory report screen

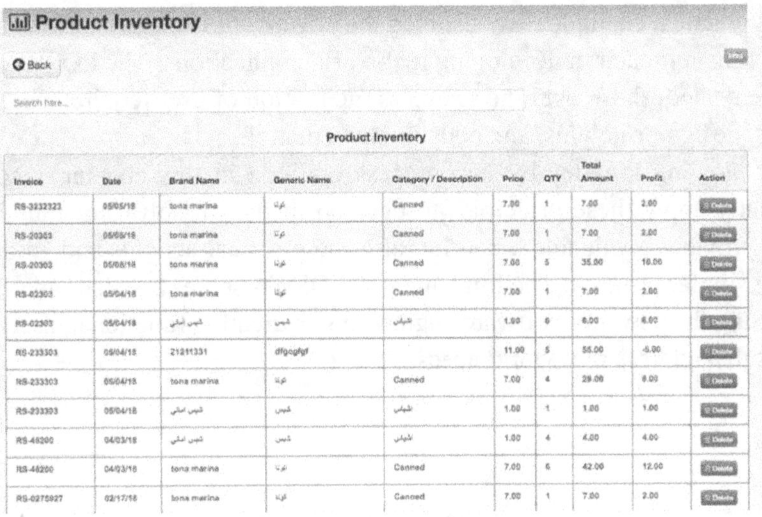

RESULTS

This chapter presented a comprehensive implementation of a Point of Sale Web Application System (POSWA) using PHP and MySQL. By following the core phases of software engineering – requirements specification, analysis, design, and implementation – we successfully transformed abstract user needs into a concrete, functional system. The implemented POSWA addresses the limitations of the existing system, providing a user-friendly, feature-rich solution tailored to the specific requirements of supermarket environments like Halhul Mall.

The chapter meticulously documented each stage of the development process. We started by analyzing the existing system, identifying its shortcomings, and defining clear functional requirements. These requirements were then translated into detailed use case diagrams, sequence diagrams, state diagrams, and class diagrams, providing a blueprint for the system's structure and behavior. Finally, we showcased the implementation details, including database design, user interface design, and key functionalities for each user role: administrator, manager, and cashier.

CONCLUSION

This practical example demonstrated the effectiveness of a systematic software engineering approach in developing real-world applications. The POSWA system, as implemented, showcases the seamless integration of user requirements, system analysis, design principles, and code implementation.

By following the step-by-step process outlined in this chapter, readers can confidently apply these principles to develop their own software solutions. The chapter provides a valuable blueprint, offering practical insights and guidance for tackling similar projects. The implemented POSWA serves as a tangible example, showcasing the power of software engineering in creating functional, user-friendly systems that address real-world needs.

Chapter 8
Deployment and Testing the Website

ABSTRACT

This chapter explains the server technologies and management services associated with the hosting and management of secure websites, server and hosting package type, website publishing, transferring the website files using the online file manager or using the FTP protocol file, search engine optimization, tools, and software used to develop websites, utilizing website technologies, tools, and techniques with good design principles to deploy a multipage website, and creating and using a test plan to review the performance and design of a multi-page website.

INTRODUCTION

Different deployment and testing methods significantly impact website performance and user experience such as using different types of Captcha for website security which involves simulating website application deployment in a pre-online environment to avoid disruptions to user visits, thus maintaining a positive user experience, Five types of Captcha used are (Tanthavech & Nimkoompai, 2019):

1. Invisible Captcha was used despite its convenience, it was not trusted by most users due to perceived insecurity.
2. reCaptcha by Google: Considered the fastest to pass among the tested Captcha types, taking only 3.09 seconds, and it is deemed the safest option.
3. Text-based Captcha: Tested as one of the Captcha types in the study but specific performance results were not provided.
4. Game-based Captcha: Tested as one of the Captcha types but specific performance results were not provided.

DOI: 10.4018/978-1-6684-4844-1.ch008

5. Math Captcha: Identified as providing the best user experience due to its entertainment value and intuitiveness, and it's considered favourable for its engaging nature.

DevOps platform used as application deployment in a pre-online environment to avoid disruptions to user visits, thus maintaining a positive user experience such as pantheon.io allows developer to build exceptional Drupal and WordPress sites with streamlined workflows, scalable infrastructure, and a lightning-fast content delivery network.

Performance testing, as discussed in (Noble, 2023), is crucial for ensuring web application functionality and performance, with open-source performance analysis tools aiding in identifying bottlenecks. Additionally, user testing methods, such as remote and traditional usability testing as explored in (Pierce et al., 2022), show similarities in user behaviour and subjective assessments, highlighting the importance of effective testing methodologies. Site-specific usability evaluations, as detailed in (Aichernig et al., 2018), help identify correctable usability concerns, ultimately improving the overall user experience of electronic health record systems.

USING LOCAL TESTING ENVIRONMENT

1.1. PHP System Deployment

This section describes how the user can install guides for your AOES system, the system will run under an Apache server environment, for testing you can use xampp server which includes php server and MySQL(West & Prettyman, 2018) server, to install the AOES with DB follow the following steps:

1. To install XAMPP on your laptop: Go to https://www.apachefriends.org/index.html
2. Use the default installation setting (XAMPP Instructions/Installation)
3. After finishing xampp installation start xampp

```
[ Store document/program files in C:/xampp/htdocs ]
```

```
[ Use "localhost/filename" to access your documents from
your browser e.g. "localhost/index.php" ]
```

4. To use XAMPP after it is installed:

Type "XAMPP" into the Search Box on your computer, then Select the XAMPP program, the XAMPP Control Panel will open as shown in Figure 1.

To start the Apache server we need to click the Apache Module Start button, in Figure 2 the Apache is running and highlighted in Green.

Figure 1. XAMPP control panel

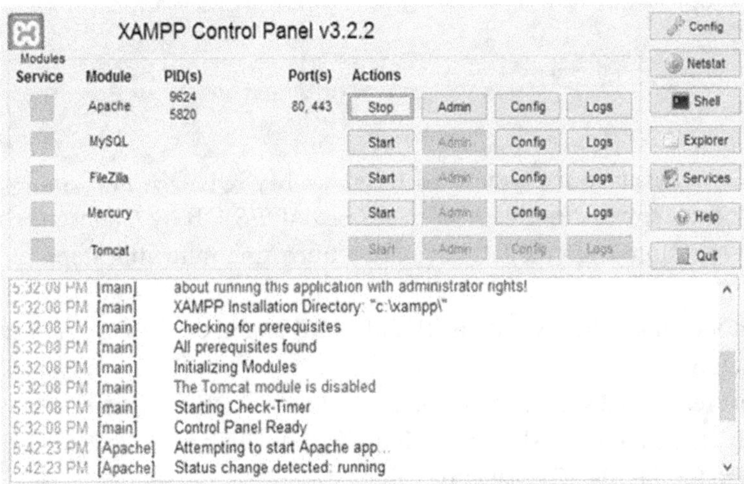

5. From deployment phase you need to create your website folder that include all .php pages in htdocs folder as shown in Figure 2, in our example we create folder name as DEC6 then we extract folder and copy dec6 folder then past folder to htdocs in File Explorer:

Figure 2. Root folder to upload .php pages

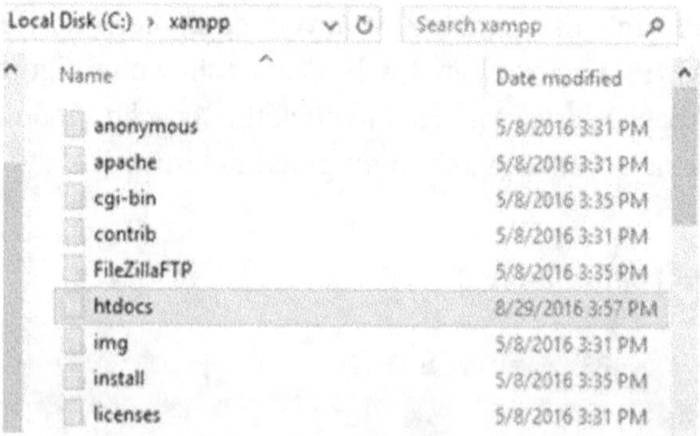

Most of website to use you need to create MySQL DB, and connect it with website, next step we describe how to restore AOES DB by Choose the database you want to restore from the left navigation tree using following step:

1. Open localhost/phpMyAdmin (Bunafit, 2017) and create DB with name such as AOESv2.
2. Open created DB such as we created previously AOESv2 then Click the Import tab to create tables as shown in Figure 3, then click the Choose File button to browse for the MySQL backup.

Figure 3. MYSQL import/restore page

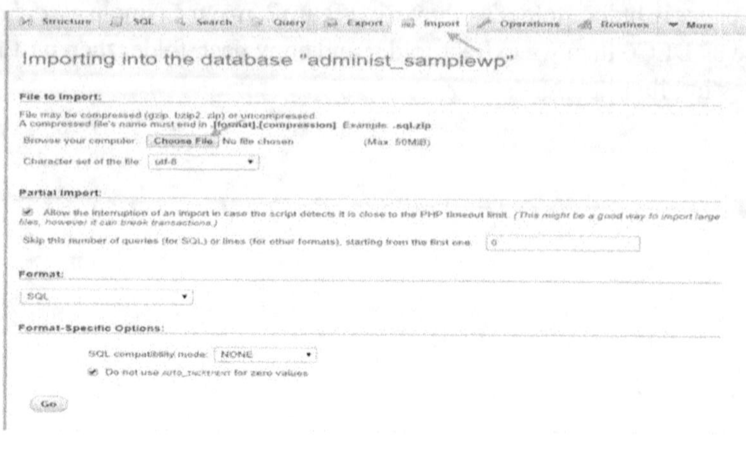

3. from the choose file select SQL file locate in project folder, then click on go button, then click Go.
4. After create DB and copy php files to htdocs, admin need to edit Connection.php file in site folder as shown in Figure 4, add username and password, add DB name as created on previous steps if different.

Figure 4. Config php connection file to include MySQL connection details

```
<?php

$con= new mysqli('localhost','root','','AOESv2')or die("Could not connect to mysql".mysqli_error($con));

?>
```

C# DEPLOYMENT PROJECT

To publish C# PWS website example that we use in previous chapter, from project folder, upload all folder files to hosting server the server should support Microsoft .net technology with internet information service (IIS) (Chitra & Satapathy, 2017). After all PWS files are uploaded and complete we need to create database name as example "morebBD" and restore the sql script from database folder in installation folder to create table and default data.

Creating a Destination Database

The Web project currently has two databases. You will deploy the contents of both databases into a single database. In a production environment, it can be more cost effective to maintain and manage one database instead of two. In the following procedure, you create the empty destination database that you will deploy website. To create a destination database

1. In Server Explorer, right-click Data Connections, and then click Create New SQL Server Database. The Create New SQL Server Database dialog box is displayed.
2. In the Server name box, enter localhost\SQLExpress.
3. In the New database name box, enter AdventureWorksTest.

When you have entered the database name, the Create New SQL Server Database dialog box resembles the following illustration, Figure 5 shown Dialog:

Figure 5. SQL server database dialog

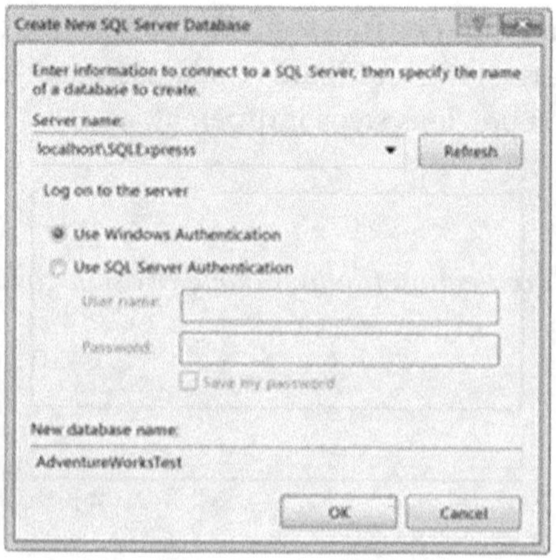

4. Click OK, the new database is created and is displayed in the Data Connections folder in the Server Explorer window, as shown on Figure 6.

Figure 6. Server explorer window

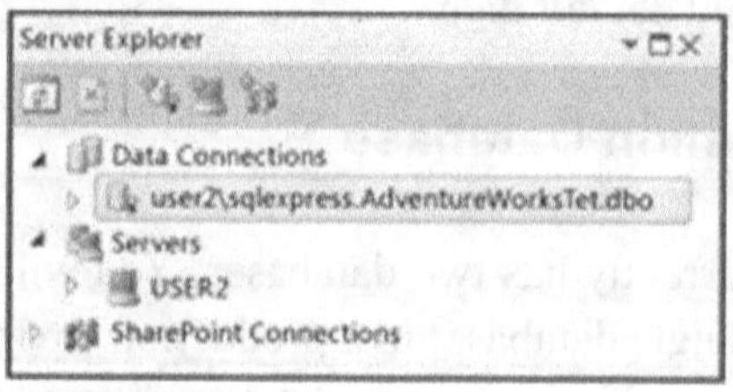

PRACTICAL EXAMPLE

As emphasized in chapter 1, establishing a domain for our project is essential. For this project, we opted for a free domain from Freenom and free hosting from byet.host. Byet.host was chosen for its attractive features, including 1GB of free space, over 51GB of bandwidth, and other capabilities that will be explored in this chapter. After creating an account and logging in to byet.host, you are greeted with a control panel. This interface empowers users to manage their websites, including file transfer and access to various features, as illustrated in the following Figure 7.

Figure 7. Control panel Bayt host server

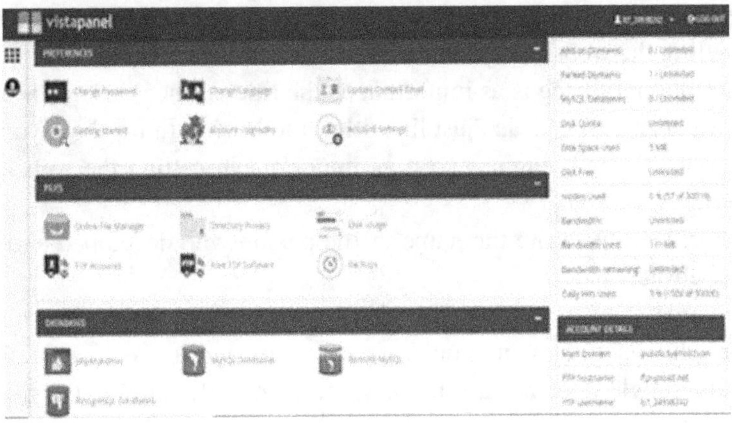

Now before transferring project files to the server host, after the complete upload all website file, we will start to declare the website to search engine, by defining the main tags in the head section as shown in Figure 8, please note that the meta tag can edit from index page, this called “seo search engine optimizer”.

Figure 8. Meta tags

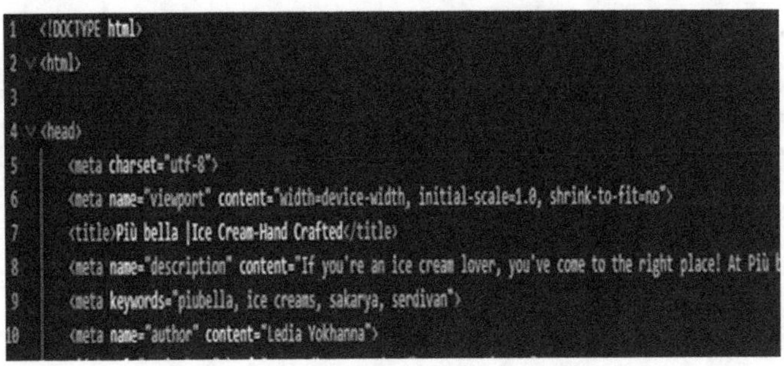

```
<!DOCTYPE html>
<html>

<head>
    <meta charset="utf-8">
    <meta name="viewport" content="width=device-width, initial-scale=1.0, shrink-to-fit=no">
    <title>Più bella |Ice Cream-Hand Crafted</title>
    <meta name="description" content="If you're an ice cream lover, you've come to the right place! At Più b
    <meta keywords="piubella, ice creams, sakarya, serdivan">
    <meta name="author" content="Ledia Yokhanna">
```

Meta tags are elements that are written in the head tag in html files, these tags do not appear to the user, but instead it is useful to the search engines actually it is key to the website because what all the search engines looks for is the content in the meta tags, and it consist of some tags which are:

- Charset: this tag defines in which character set a website is coded.
- Viewport: here it allows the control of the expanding and shrinking of the website in different devices small or large.
- Title: this tag is mostly the most important of all meta tags since it defines the title of the website that will be shown above in the browser's tab, unlike other tags that are invisible this tag is available to all users and it is the main reason of getting the website a good rank.
- Description: this tag is as important as the title is since this one is also available to the users, and also just like titles it tells everything about your website is about to search engines so it as important in getting the website a good ranking.
- Author: here we write the name of the person who developed and designed this website.
- Keywords: they are some keywords that we decide that describes our website or in which situations it should appear, years ago it was beneficial but now mostly google does not use the keywords and rely only on the title and the description.

To raise the site to the first levels of search sites there are many online tools can help to achieve this task which give us keywords that assist us with finding the words utilized in each nation through which we can know the most significant things that clients around the globe are looking for and that suit our organization administrations, using keywordtool.io allow us to pick the most utilized keyword to add to our website to for SEO as show in Figure 9.

Figure 9. keywordtool

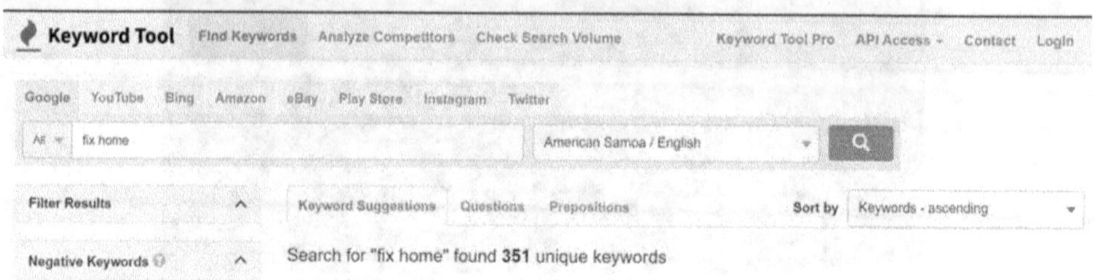

We take the keywords that we got from the Keyword Tool website and add them in the codes inside the HTML specifically within the head tag. Because the user is practically writing a keyword in the search process, the search engine searches for this word within the head tag.

Now after all meta tags are defined the we site ready transfer the website files, and we have two ways to transfer the files to the server which are:

1. online file manager:

We choose the online file manager form the control panel, then transfer local files to the htdocs folder, by drag and drop the files to the htdocs, the result will be as shown in next Figure.

Figure 10. Htdocs folder

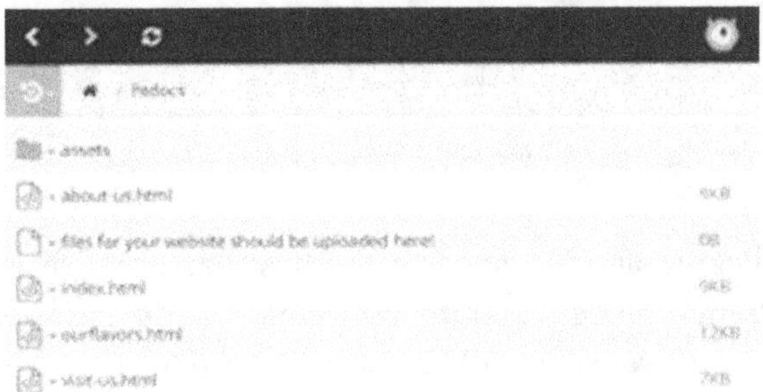

2. Using the ftp Protocol file.

This method which is more guaranteed used by by drag and drop the files, it work based on ftp protocol that is offered from the byet.host website, to upload the files we can use any ftp client software, and in this chapter we sue FileZilla as it's free software, user can install local computer then open it as show in Figure 11.

Figure 11. FileZilla main screen

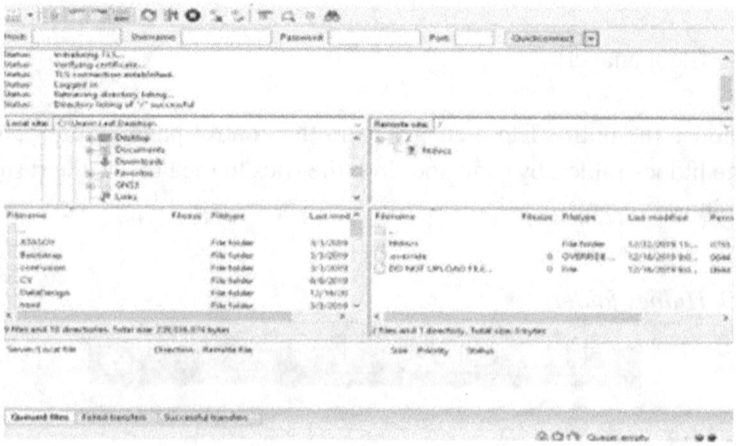

After open filezilla and by entering the host name which is provided to me during the registration process and the username and the password, then you will be accessed to this software and will be able to upload the files from your computer to server, as shown above in our example we choose the file that we want to upload to form the desktop on the left side, to upload just drag and drop the files into the htdocs, and the beautiful thing about file zilla is that below it shows the failed and successfully transferred files, which is more guaranteed.

3. Configure domain server with hosting server

first, from hosting server which describe in previous step open the control panel of the server and from the domains section chose aliases (parked domains) link, then entering your domain that register, in our example we use freenom domains as shown in Figure 12, after entering in the domain name section you should back to the main page on freenom.com and add DNS server listed from hosting server as showing in our example ns1.byte.org, chose the nameserver option out of management tools.

Figure 12. Configure aliases domain

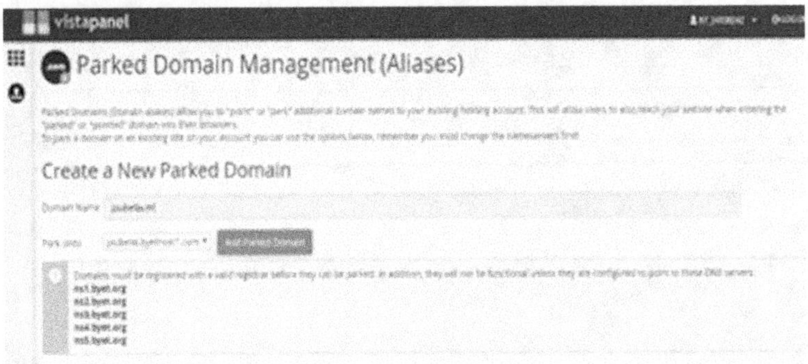

CONNECTING THE CEO TO GOOGLE ANALYTICS

After we published the website we need Analytics and statistical info about user how visitor the site, see how clients are collaborating with your site and applications, what page thy open, time need to load to the client browser, what type of devices thy use, user browser type, where the user is, the number of users, type of ip used, and which screen resolution client use, this info help us to next phase of development in software engineering, many tools available and most use is Google Analytics we can use as advantages and favorable circumstances in building up the site analytics causes you see how individuals utilize your locales and applications, so you can make a move to improve their experience, to add website we need to create account using google analytics will take these steps:

1. going to googleanalytics.com
2. click on set up for free button
3. choose an account name, and press next
4. selecting the web since you want to connect (we use domain name that we used in previous example) then also click on next, after registering our website custom in analytics, we have copied and pasted the code into html files (index file), then after you add code Google Analytics will verify the page and if code found the connect step completed as shown in Figure 13.

Figure 13. Analytics cod

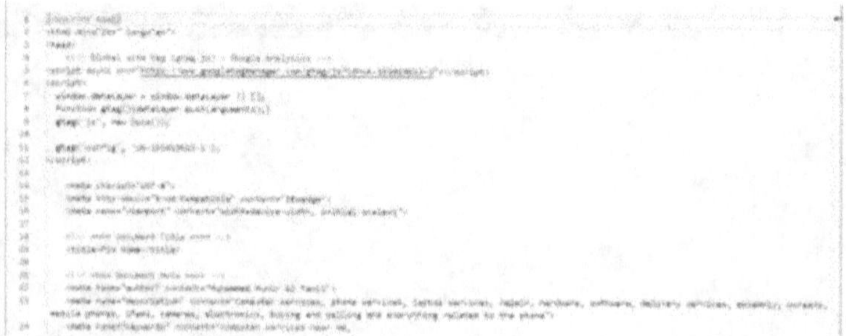

Next step you can visit google analytic main page which include all statistical info appearance of users who use our site as shown in Figure 14.

Figure 14. Analytics visitor

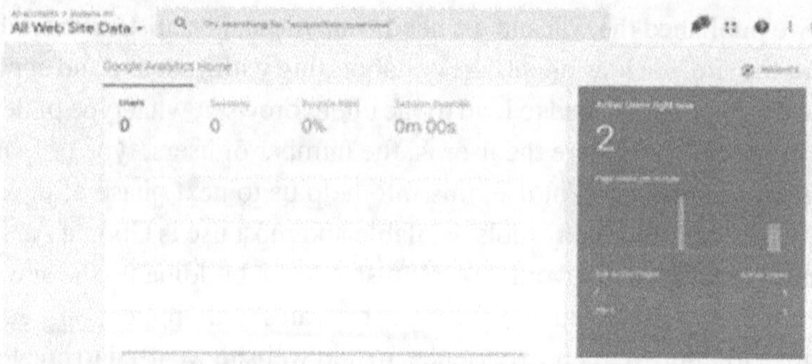

Web Testing Checklists

1) Functionality Testing
    - Check all the links
    - Test forms on all pages
    - Cookies Testing
    - Validate your HTML/CSS
    - Database Testing
2) Usability testing

Usability testing is a crucial step in software development, evaluating a system's ease of use and identifying areas for improvement. It involves observing real users interacting with the system and gathering feedback on their experience. The goal is to ensure that the system is intuitive, efficient, and enjoyable to use, this step include the following criteria:

- Test for Navigation
- The website should be easy to use.
- The instructions provided should be very clear.
- Check if the instructions provided are perfect to satisfy its purpose.
- The main menu should be provided on each page.
- It should be consistent enough.
- Content Checking

To test and check our website you can use many tools such as Pingdom website, as shown in Figure 15.

Figure 15. Check our website

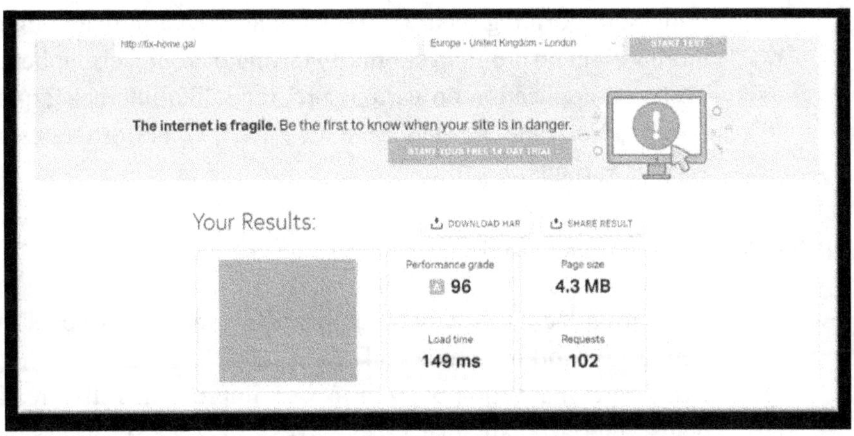

3) Interface testing

main interfaces are:

- Web server and application server interface
- Application server and Database server interface

4) Compatibility testing

The similarity of your site is a significant testing perspective. See which similarity test to be executed:

- Program similarity
- Working framework similarity
- Versatile perusing
- Printing choices

Program Compatibility:

- In my web-testing profession, we have encountered this as the most affecting piece of site testing.
- A few applications are reliant on programs. Various programs have various arrangements and settings that your website page ought to be good with.
- Your site coding ought to be a cross-program stage good. In the event that you are utilizing java contents or AJAX calls for UI usefulness, performing security checks or approvals at that point give more weight on program similarity testing of your web application.
- Test web application on various programs like Internet Explorer, Firefox, Netscape Navigator, AOL, Safari, Opera programs with various variants.

Operating system Compatibility:

- Some usefulness in your web application is that it may not be good with every working framework.
- Every new innovation utilized in web improvement like realistic plans, interface calls like various API's may not be accessible in every single Operating System.
- Consequently, test your web application on various working frameworks like Windows, Unix, MAC, Linux, Solaris with various OS flavors.

Portable Browsing:

- We are in the new innovation time. So, in future Mobile perusing will shake. Test your pages on portable programs. Similarity issues might be there on cell phones also.

Printing Options:

- On the off chance that you are giving page-printing alternatives, at that point ensure textual styles, page arrangement, page designs, and so on., are getting printed appropriately. Pages should fit the paper size or according to the size referenced in the printing alternative.

5) Performance testing

The web application should sustain to heavy load. Web performance testing should include:

- Web Load Testing
- Web Stress Testing
- Connection Speed

6) Security testing

Following are some of the test cases for web security testing:

- Test by pasting the internal URL directly into the browser address bar without login. Internal pages should not open.
- If you are logged in using username and password and browsing internal pages then try changing URL options directly. I.e. If you are checking some publisher site statistics with publisher site ID= 123. Try directly changing the URL site ID parameter to different site ID which is not related to the logged-in user. Access should be denied for this user to view other's stats.
- Try some invalid inputs in input fields like login username, password, input text boxes, etc. Check the system's reaction to all invalid inputs.
- Web directories or files should not be accessible directly unless they are given download option.

- Test the CAPTCHA for automating script logins.
- Test if SSL is used for security measures. If it is used, the proper message should get displayed when users switch from non-secure HTTP:// pages to secure HTTPS:// pages and vice versa.
- All transactions, error messages, security breach attempts should get logged in log files somewhere on the webserver.

Information consistency is likewise significant in a web application. Check for information respectability and mistakes while you alter, erase, change the structures or do any Database Testing related usefulness. Check if all the database questions are executing effectively, information is recovered and furthermore refreshed accurately. More on database testing could be a heap on DB, we will address this in web burden or execution testing beneath. In testing the usefulness of the sites, the accompanying ought to be tried:
Connections

I. Inward Links

ii. Outer Links
iii. Mail Links
iv. Broken Links

Structures

I. Field approval

ii. Mistake message for wrong info
iii. Discretionary and Mandatory fields

Performance Testing

The web application ought to support to overwhelming burden. Web execution testing ought to include Web Load Testing or Web Stress Testing which used to test application execution on various web association speeds as describe in following potins:

Web Load Testing: You have to test if numerous clients are getting to or mentioning a similar page. Could the framework support in top burden times? The site should deal with numerous synchronous client demands, huge info information from clients, concurrent association with DB, substantial burden on explicit pages, and so forth.

Web Stress Testing: Generally, stress implies extending the framework past its predefined limits. Web pressure testing is performed to break the webpage by giving pressure and it's checked concerning how the framework responds to pressure and how it recoups from crashes. Stress is commonly given on input fields, login and join zones. In web execution, testing site usefulness on various working frameworks and distinctive equipment stages is checked for programming and equipment memory spillage blunders.

Execution testing can be applied to comprehend the site's versatility or to benchmark the exhibition in the earth of outsider items, for example, servers and middleware for potential buy.

Association Speed

Tried on different systems like Dial-Up, ISDN, and so forth.

Load

I. What is the no. of clients per time?

ii. Check for top loads and how the framework acts

iii. A lot of information got to by the client

Stress

I. Constant Load

ii. Execution of memory, CPU, document dealing with, and so on.

Security Testing

For web security analysis using penetration testing developer can implementing the Open Web Application Security Project (OWASP) which involves conducting legal testing activities that simulate hacker behavior to detect vulnerabilities in web applications (Riandhanu, 2022).

Test by sticking the inner URL straightforwardly into the program address bar without login. On the off chance that you are signed in utilizing username and secret phrase and perusing interior pages at that point take a stab at changing URL choices legitimately. For example, In the event that you are checking some distributer site measurements with distributer site ID= 123. Attempt legitimately changing the URL site ID parameter to various site ID which isn't identified with the signed in client. Access ought to be denied for this client to see other's details.

Quality assurance encompasses a broad spectrum of activities aimed at measuring and enhancing the quality of a product. This includes every stage of the development process, from training and team preparation to the final product release. In contrast, quality control focuses specifically on verifying the product's quality. It involves detecting and rectifying faults or defects before the product is released.

Essentially, every software program possesses two fundamental aspects: a static structure, which refers to the organization of its source code, and a dynamic behavior, which describes how it functions during execution. However, intermediate software products, such as requirements documents, analysis reports, and design specifications, only exhibit a static structure as they are primarily documents and not executable code (Tsui et al., n.d.).

RESULTS

This chapter delves into the crucial aspects of deploying and testing a web application, building upon the analysis, design, and implementation phases discussed in previous chapters. Having established a thorough understanding of system requirements, user interactions, and architectural design, we now focus on translating those blueprints into a tangible, functional system.

A practical website example was successfully deployed on the byet.host server, all functionalities, including user registration, login, exam creation, and result retrieval, were tested and confirmed to be working as expected. Website demonstrated good performance under load testing, handling multiple concurrent user requests. Security testing revealed no major vulnerabilities, and appropriate measures were implemented to address minor issues. Usability testing provided valuable feedback, leading to improvements in navigation, instructions, and the overall user interface.

CONCLUSION

This chapter highlighted the critical steps involved in deploying and testing a web application, emphasizing the importance of a thorough and systematic approach. By following the guidelines presented, developers can ensure a successful transition from a development environment to a live, production-ready system. The practical examples and testing methodologies discussed in this chapter equip readers with the knowledge and tools necessary to confidently deploy and evaluate their own web applications.

REFERENCES

Aichernig, B. K., Kann, S., & Schumi, R. (2018). Statistical Model Checking of Response Times for Different System Deployments. *Lecture Notes in Computer Science (Including Subseries Lecture Notes in Artificial Intelligence and Lecture Notes in Bioinformatics), 10998 LNCS*, 153–169. 10.1007/978-3-319-99933-3_11

Bunafit, N. (2017). *Dasar Pemrograman Web PHP–MySQL Dengan Dreamweaver*. GavaMedia.

Chitra, L. P., & Satapathy, R. (2017). Performance comparison and evaluation of Node.js and traditional web server (IIS). *2017 International Conference on Algorithms, Methodology, Models and Applications in Emerging Technologies, ICAMMAET 2017, 2017-January*. 10.1109/ICAMMAET.2017.8186633

Noble, E. (2023). Deployment. *Pro T-SQL*, 2022, 321–345. 10.1007/978-1-4842-9256-3_12

Pierce, R. P., Eskridge, B. R., Ross, B., Day, M. A., Dean, B., & Belden, J. L. (2022). Improving the User Experience with Discount Site-Specific User Testing. *Applied Clinical Informatics*, 13(5), 1040–1052. 10.1055/s-0042-175822236323335

Riandhanu, I. O. (2022). *Analisis Metode Open Web Application Security Project (OWASP) Menggunakan Penetration Testing pada Keamanan Website Absensi*. Jurnal Informasi Dan Teknologi. 10.37034/jidt.v4i3.236

Tanthavech, N., & Nimkoompai, A. (2019). CAPTCHA: Impact of Website Security on User Experience. *ACM International Conference Proceeding Series, Part F147957*, 37–41. 10.1145/3321454.3321459

Tsui, F. F., Karam, O., & Bernal, B. (n.d.). *Essentials of software engineering*. Retrieved May 24, 2024, from https://www.jblearning.com/catalog

West, A. W., & Prettyman, S. (2018). *Practical PHP 7, MySQL 8, and MariaDB Website Databases*. 10.1007/978-1-4842-3843-1

Compilation of References

About the Unified Modeling Language Specification Version 2.0. (n.d.). Retrieved May 24, 2024, from https://www.omg.org/spec/UML/2.0/

Aichernig, B. K., Kann, S., & Schumi, R. (2018). Statistical Model Checking of Response Times for Different System Deployments. *Lecture Notes in Computer Science (Including Subseries Lecture Notes in Artificial Intelligence and Lecture Notes in Bioinformatics), 10998 LNCS*, 153–169. 10.1007/978-3-319-99933-3_11

Alter, S. (2001). Which Life Cycle—Work System, Information System, or Software? *Communications of the Association for Information Systems*, 7. Advance online publication. 10.17705/1CAIS.00717

Apple Inc. (2017). Xcode IDE. In *Apple Developer*. Author.

Aziz, B. (2020). Modeling and Analyzing an Industry 4.0 Communication Protocol. *IEEE Internet of Things Journal*, 7(10), 10120–10127. Advance online publication. 10.1109/JIOT.2020.2999325

Baiskar, Y. (2022). MERN: A Full-Stack Development. *International Journal for Research in Applied Science and Engineering Technology*, 10(1), 1029–1035. Advance online publication. 10.22214/ijraset.2022.39982

Barrett, K., Hunt, C. J., Lange, L., & Meyer, A. S. (2020). Conserved unique peptide patterns (CUPP) online platform: Peptide-based functional annotation of carbohydrate active enzymes. *Nucleic Acids Research*, 48(W1), W110–W115. Advance online publication. 10.1093/nar/gkaa37532406917

Bell, C. P. M., Rees, J. M., Naylor, T., Mayne, N. J., Jeffries, R. D., Mamajek, E. E., & Rowe, J. (2014). Pre-main-sequence isochrones - III. The cluster collaboration isochrone server. *Monthly Notices of the Royal Astronomical Society*, 445(4), 3496–3511. Advance online publication. 10.1093/mnras/stu1944

Benfell, A. (2021). Modeling functional requirements using tacit knowledge: A design science research methodology informed approach. *Requirements Engineering*, 26(1), 25–42. Advance online publication. 10.1007/s00766-020-00330-4

Brereton, P., Kitchenham, B. A., Budgen, D., Turner, M., & Khalil, M. (2007). Lessons from applying the systematic literature review process within the software engineering domain. *Journal of Systems and Software*, 80(4), 571–583. Advance online publication. 10.1016/j.jss.2006.07.009

Brownlee, N., Claffy, K. C., & Nemeth, E. (2001). DNS measurements at a root server. *Conference Record / IEEE Global Telecommunications Conference, 3.* 10.1109/GLOCOM.2001.965864

Bunafit, N. (2017). *Dasar Pemrograman Web PHP–MySQL Dengan Dreamweaver*. GavaMedia.

Campbell, S. (2021). *Flask vs Django: What's the Difference Between Flask & Django?* Guru99.

Casteren, W. van. (2017). The Waterfall Model And The Agile Methodologies : A Comparison By Project Characteristics-Short The Waterfall Model and Agile Methodologies. *Academic Competences in the Bachelor*.

Chitra, L. P., & Satapathy, R. (2017). Performance comparison and evaluation of Node.js and traditional web server (IIS). *2017 International Conference on Algorithms, Methodology, Models and Applications in Emerging Technologies, ICAMMAET 2017, 2017-January*. 10.1109/ICAMMAET.2017.8186633

Christanto, F. W., & Sani Suprayogi, M. (2018). Enhancement Network Monitoring System Functionality by Implementing an Android-based Notification System to Monitor Virtual Servers on Cloud Computing Network. *IJAIT*, 2(1), 7. Advance online publication. 10.25124/ijait.v2i01.1002

Chu, W. C., Lu, C. W., Yang, H., & He, X. (2000). A formal approach for component retrieval and integration analysis. *Journal of Software Maintenance: Research and Practice*, 12(6), 325–342. Advance online publication. 10.1002/1096-908X(200011/12)12:6<325::AID-SMR217>3.0.CO;2-#

Cloudflare. (2020). *What Is DNS? | How DNS Works | Cloudflare*. Cloudflare Inc.

Cortez, R., & Vazhenin, A. (2015). Virtual model-view-controller design pattern: Extended MVC for service-oriented architecture. *IEEJ Transactions on Electrical and Electronic Engineering*, 10(4), 411–422. Advance online publication. 10.1002/tee.22101

D'Arcy, H. (2009). *IntelliJ IDEA*. RefCardz.

Davis, W. S. (2019). Data flow diagrams. *The Information System Consultant's Handbook*, 175–188. 10.1201/9781420049107-24

Dickerson, C. E., Roslan, R., & Ji, S. (2018). A Formal Transformation Method for Automated Fault Tree Generation from a UML Activity Model. *IEEE Transactions on Reliability*, 67(3), 1219–1236. Advance online publication. 10.1109/TR.2018.2849013

Duarte, P., Costa, C. J., & Costa, P. (2008). *Webstorm*. 10.1145/1456536.1456592

Dziallas, M. (2020). How to evaluate innovative ideas and concepts at the front-end?: A front-end perspective of the automotive innovation process. *Journal of Business Research*, 110, 502–518. Advance online publication. 10.1016/j.jbusres.2018.05.008

Eazley, D., Jones, B. K., Martelli, A., Ravenscroft, A., Holden, S., Schlusser, T., Reitz, K., Stephenson, B., & Dwyer, G. (2017). The Hitchhiker's Guide to Python – Best Practices for Development. In *Packt Publishing* (Vol. 44, Issue 8).

Examino - Online Exam Maker. (n.d.). Retrieved December 9, 2022, from https://examino .finicode.com/ ISO/IEC

Express.js. (2017). *Express - Node.js web application framework*. Expressjs.Com.

Firmansyah Agustian. (2017). Adobe Dreamweaver. *Jurnal Desain Web, 1*.

Firmansyah, A. S., & Riadi, I. (2014). Analisis dan perancangan proxy server menggunakan virtual machine. *Sarjana Teknik Informatika*.

Froehlich, A., Rosencrance, L., & Gattine, K. (2021). What is the OSI model? The 7 layers of OSI explained. *Tech Target*.

Garais, G. E., & Enaceanu, A.-S. (2016). Open Source Servers and Website Platforms Security. *Journal of Information Systems & Operations Management*.

Haris, N. A., & Hasim, N. (2019). PHP frameworks usability in web application development. *International Journal of Recent Technology and Engineering, 8*(3). 10.35940/ijrte.C1020.1083S19

Heemstra, F. J. (1992). Software cost estimation. *Information and Software Technology*, 34(10), 627–639. Advance online publication. 10.1016/0950-5849(92)90068-Z

Hilburn, T. B., Hirmanpour, I., Khajenoori, S., Turner, R., & Qasem, A. (1999). *A Software Engineering Body of Knowledge Version 1.0*. Academic Press.

How To Become A DevOps Engineer In 2022: A Comprehensive Guide. (n.d.). Retrieved December 8, 2022, from https://devopscube.com/become-devops-engineer/

Ikhsan, A., Ma, S., Perdananto, A., Studi Magister Informatika, P., Eresha, S., Selatan, T., & Raya Puspitek No, J. (2020). Aplikasi Pemesanan Tiket Seminar Secara Online Berbasis Web Di Universitas Pamulang. *Jurnal Teknologi Informasi ESIT*.

Jang, J. W., Verbeek, F., & Ravindran, B. (2021). Verification of Functional Correctness of Code Diversification Techniques. *Lecture Notes in Computer Science (Including Subseries Lecture Notes in Artificial Intelligence and Lecture Notes in Bioinformatics), 12673 LNCS*, 160–179. 10.1007/978-3-030-76384-8_11

Khan, F., Kumar, R. L., Kadry, S., & Nam, Y. (2021). The future of software engineering: Visions of 2025 and beyond. *Iranian Journal of Electrical and Computer Engineering*, 11(4), 3443. Advance online publication. 10.11591/ijece.v11i4.pp3443-3450

Kumar, V., & Roy, O. P. (2021). Security and challenges in voice over internet protocols: A survey. *IOP Conference Series. Materials Science and Engineering*, 1020(1), 012020. Advance online publication. 10.1088/1757-899X/1020/1/012020

Ľachová, K., & Trebuňa, P. (2019). Modeling of electronic kanban system by using entity relationship diagrams. *Acta Logistica*, 6(3), 63–66. Advance online publication. 10.22306/al.v6i3.115

Leite, L., Rocha, C., Kon, F., Milojicic, D., & Meirelles, P. (2019). A survey of DevOps concepts and challenges. In *ACM Computing Surveys* (Vol. 52, Issue 6). 10.1145/3359981

Mahler, T. (2014). A gTLD right? Conceptual challenges in the expanding internet domain namespace. *International Journal of Law and Information Technology*, 22(1), 27–48. Advance online publication. 10.1093/ijlit/eat018

Martínez, Y. V., Domínguez, E. L., Velázquez, Y. H., Isidro, S. D., Medina Nieto, M. A., & de La Calleja, J. (2019). Layered software architecture for the development of third-generation video surveillance systems. *IEEE Access : Practical Innovations, Open Solutions*, 7, 98507–98521. Advance online publication. 10.1109/ACCESS.2019.2930401

McGillivray, K. (2014). Give it away now? Renewal of the IANA functions contract and its role in internet governance. *International Journal of Law and Information Technology*, 22(1), 3–26. Advance online publication. 10.1093/ijlit/eat017

Moreb, M. (2022). *Practical Forensic Analysis of Artifacts on iOS and Android Devices*. Practical Forensic Analysis of Artifacts on IOS and Android Devices. 10.1007/978-1-4842-8026-3

Moreb, M., Mohammed, T. A., & Bayat, O. (2020). A Novel Software Engineering Approach Toward Using Machine Learning for Improving the Efficiency of Health Systems. *IEEE Access : Practical Innovations, Open Solutions*, 8, 23169–23178. 10.1109/ACCESS.2020.2970178

Nawaz, A., Arora, A. S., Ali, W., Saxena, N., Khan, M. S., Yun, C. M., & Lee, M. (2022). Intelligent Human-Machine Interface: An Agile Operation and Decision Support for an ANAMMOX SBR System at a Pilot-Scale Wastewater Treatment Plant. *IEEE Transactions on Industrial Informatics*, 18(9), 6224–6232. Advance online publication. 10.1109/TII.2022.3153468

NetBeans. (2015). NetBeans IDE. *Features, Jsr 299*.

Noble, E. (2023). Deployment. *Pro T-SQL*, 2022, 321–345. 10.1007/978-1-4842-9256-3_12

Oliveira, L., Rodrigues, J. J. P. C., Kozlov, S. A., Rabêlo, R. A. L., & de Albuquerque, V. H. C. (2019). MAC layer protocols for internet of things: A survey. In *Future Internet* (Vol. 11, Issue 1). 10.3390/fi11010016

Paolone, G., Marinelli, M., Paesani, R., & di Felice, P. (2020). Automatic code generation of mvc web applications. *Computers*, 9(3), 56. Advance online publication. 10.3390/computers9030056

Petrikoglou, A., & Kaskalis, T. H. (2019). Full stack web development teaching: Current status and a new proposal. *WEBIST 2019 - Proceedings of the 15th International Conference on Web Information Systems and Technologies*. 10.5220/0008066202180225

Pierce, R. P., Eskridge, B. R., Ross, B., Day, M. A., Dean, B., & Belden, J. L. (2022). Improving the User Experience with Discount Site-Specific User Testing. *Applied Clinical Informatics*, 13(5), 1040–1052. 10.1055/s-0042-175822236323335

Raja, J. Z., Chakkol, M., Johnson, M., & Beltagui, A. (2018). Organizing for servitization: Examining front- and back-end design configurations. *International Journal of Operations & Production Management*, 38(1), 249–271. Advance online publication. 10.1108/IJOPM-03-2016-0139

Riandhanu, I. O. (2022). *Analisis Metode Open Web Application Security Project (OWASP) Menggunakan Penetration Testing pada Keamanan Website Absensi.* Jurnal Informasi Dan Teknologi. 10.37034/jidt.v4i3.236

Salas-Zárate, M. D. P., Alor-Hernández, G., Valencia-García, R., Rodríguez-Mazahua, L., Rodríguez-González, A., & López Cuadrado, J. L. (2015). Analyzing best practices on Web development frameworks: The lift approach. In *Science of Computer Programming* (Vol. 102). 10.1016/j.scico.2014.12.004

Shankar, P., Morkos, B., Yadav, D., & Summers, J. D. (2020). Towards the formalization of non-functional requirements in conceptual design. *Research in Engineering Design*, 31(4), 449–469. Advance online publication. 10.1007/s00163-020-00345-6

Syarifudin, A. (2020). *Konfigurasi DHCP Server.* Fakultas Komputer.

Tanthavech, N., & Nimkoompai, A. (2019). CAPTCHA: Impact of Website Security on User Experience. *ACM International Conference Proceeding Series, Part F147957*, 37–41. 10.1145/3321454.3321459

Tiwari, S., & Singh Rathore, S. (2019). Teaching Software Process Models to Software Engineering Students: An Exploratory Study. *Proceedings - Asia-Pacific Software Engineering Conference, APSEC*. 10.1109/APSEC48747.2019.00049

Tsui, F. F., Karam, O., & Bernal, B. (n.d.). *Essentials of software engineering.* Retrieved May 24, 2024, from https://www.jblearning.com/catalog

Turay, B. (2021). Analysis of Seven Layered Architecture of Osi Model. SSRN *Electronic Journal.* 10.2139/ssrn.3815237

Tutorials Point. (2017). *Bootstrap responsive web development.* Tutorials Point.

USATestprep: K-12 standards-aligned practice tests. (n.d.). Retrieved December 9, 2022, from https://www.usatestprep.com/

Vishwas Kotwal, D., Rajendra Bhadke, S., Sanjay Gunjal, A., & Biswas, P. (2015). Online examination system. *International Research Journal of Engineering and Technology*. www.IJCSI.org

Weinberg, J. (2000). ICANN and the problem of legitimacy. *Duke Law Journal*, 50(1), 187. Advance online publication. 10.2307/1373114

West, A. W., & Prettyman, S. (2018). *Practical PHP 7, MySQL 8, and MariaDB Website Databases*. 10.1007/978-1-4842-3843-1

Wilkowski, M. (2021). Was this the real Web? Quantitative overview of the Polish... ccTLD Internet Archive data (1996–2001). *Archeion*, 122, 44–68. Advance online publication. 10.4467/26581264ARC.21.015.14495

Williams, T., Vo, H., Samset, K., & Edkins, A. (2019). The front-end of projects: A systematic literature review and structuring. *Production Planning and Control*, 30(14), 1137–1169. Advance online publication. 10.1080/09537287.2019.1594429

Wulandari, H., Suherman, S., & Razali, R. (2022). Pelatihan administrasi sistem jaringan berbasis windows pada guru dan siswa (Studi kasus: Sekolah smk madani marendal I). *RESWARA. Jurnal Pengabdian Kepada Masyarakat*, 3(1), 185–190. Advance online publication. 10.46576/rjpkm.v3i1.1637

Xian, F., Srisa-an, W., & Jiang, H. (2008). Garbage collection: Java application servers' Achilles heel. *Science of Computer Programming*, 70(2–3), 89–110. Advance online publication. 10.1016/j.scico.2007.07.008

Xu, H., Zhang, Z., Han, B., & Yan, J. (2021). Identifying Influential SLD Authoritative Name Servers on the Internet. *Frontiers in Physics (Lausanne)*, 9, 768205. Advance online publication. 10.3389/fphy.2021.768205

Yule, P., & Cooper, R. P. (2003). Express: A Web-based technology to support human and computational experimentation. *Behavior Research Methods, Instruments, & Computers*, 35(4), 605–613. Advance online publication. 10.3758/BF0319554014748505

Zammetti, F. (2020). Modern Full-Stack Development. In *Modern Full-Stack Development*. 10.1007/978-1-4842-5738-8

Zhang, F., Sun, G., Zheng, B., & Dong, L. (2021). Design and implementation of energy management system based on spring boot framework. *Information (Basel)*, 12(11), 457. Advance online publication. 10.3390/info12110457

Zhang, H., Liu, W., Xiong, H., & Dong, X. (2018). Analyzing data flow diagrams by combination of formal methods and visualization techniques. *Journal of Visual Languages and Computing*, 48, 41–51. Advance online publication. 10.1016/j.jvlc.2018.08.001

Zimmermann, H. (1980). OSI Reference Model-The ISO Model of Architecture for Open Systems Interconnection. *IRE Transactions on Communications Systems*, 28(4), 425–432. Advance online publication. 10.1109/TCOM.1980.1094702

Zou, L., Wang, Z., Hu, J., Liu, Y., & Liu, X. (2021). Communication-protocol-based analysis and synthesis of networked systems: Progress, prospects and challenges. *International Journal of Systems Science*, 52(14), 3013–3034. Advance online publication. 10.1080/00207721.2021.1917721

About the Author

Mohammed Adnan Moreb is a Vice President of Academic Affairs at Smart University College for Modern Education. He obtained his Ph.D. in Electrical and Computer Engineering. Dr. Moreb Expertise in Cybercrimes and Digital Evidence Analysis, specifically focusing on Information and Network Security, with a strong publication track record, work for both conceptual and practical which built during works as a system developer and administrator for the data center for more than 10 years, config, install, and admin enterprise system related to all security configuration, he improved his academic path with the international certificate such as CCNA, MCAD, MCSE; Academically he teaches the graduate-level courses such as Information and Network Security course, Mobile Forensics course, Advanced Research Methods, Computer Network Analysis and Design, and Artificial Intelligence Strategy for Business Leaders. Dr. Moreb founded a new framework and methodology specialized in software engineering for machine learning in health informatics named SEMLHI which investigates the interaction between software engineering and machine learning within the context of health systems. The SEMLHI framework includes four modules (software, machine learning, machine learning algorithms, and health informatics data) that organize the tasks in the framework using a SEMLHI methodology, thereby enabling researchers and developers to analyze health informatics software from an engineering perspective and providing developers with a new road map for designing health applications with system functions and software implementations.

Index

U

W

www.ingramcontent.com/pod-product-compliance
Lightning Source LLC
LaVergne TN
LVHW061541290926
849162LV00004B/874